Irrationality in Nature
or in Science ?

Probing a rational energy and mind world

Helmut Tributsch

Dedicated to observers of science,

who still try to rely on their innate common sense.

CONTENT

ORIGIN AND CHALLENGE OF IRRATIONALITY..1
  1 Irrationality in physics and in society 1
  2 Perception and non-perception of scientific phenomena..................7
  3 Perception with contradictory symbolic forms?.............................16
  4 How to deal with irrationality 18
  5 Two kinds of energy properties 20
  6 Energy dilution in space and consequences....................................24

DYNAMIC ENERGY: PARADIGM CHANGE TOWARDS RATIONALITY 31
  7 A dynamic energy concept: why? 31
  8 Observations on entropic time orientation and information 36
  9 The new energy theorem...39
  10 Decrease of energy per state and principle of least action 43
  11 Thermodynamic laws are consequences...46
  12 How did quantum physics become irrational?................................52
  13 How can quantum irrationality be overcome? 55
  14 What does quantum uncertainty mean in nature? 64
  15 Paradoxes can be eliminated 67
  16 What can be said about non-locality in general?...........................69

RETURNING TO A CREDIBLE COSMOLOGY..73
  17 The ever constant light velocity: what does it really mean? 73
  18 The four-dimensional space-time: an illusion 79
  19 Time travelling: endless paradoxes..91
  20 Elementary particles and fundamental forces 94
  21 Gravitation: information self-image of matter..............................100
  22 The cosmological redshift: a new interpretation 109
  23 The Big Bang irrationality..120
  24 Cosmological objects from galaxies to black holes 125

TIME, EVOLUTION AND MIND 134
  25 An alternative brief history of time 134
  26 The time arrow and self-organization of matter 138
  27 Evolution: the aim is high energy turnover and mind 142
  28 Self-organization of information: consciousness...........................147
  29 The genetic code self-organized 156
  30 Man and the evolution of the universe 160

31 Mankind's energy challenge: towards success or disaster?............................169
32 Money as an equivalent of energy..171
33 What is the equivalent of stockbroking and financial economy?.................173
34 Irrationalities that could be eliminated from science theories......................175
35 Evolution including mind...180

A RATIONALLY UNDERSTANDABLE NATURE 184
36 What could be learned from a dynamic energy world?.................................184
37 Concluding remarks 204

References 206

Abstracts.. 216

PREFACE

This book was written, because I feel that science, with all its relevance and its successes in modern life, as well as the fascination it creates, has also got on the wrong track. It is a track which has significant sociological and educational consequences. By starting to admit irrational scientific concepts and conclusions via quantum theory and the theory of relativity one century ago and by applying them to many natural phenomena, an environment of tolerance of "non-understanding" was created. It now penetrates science as a network of established theories which contradict human logic. The abandonment of human logic may prevent us from solving logic problems and also from understanding that a convincing concept of nature has to include an explanation of intelligence and consciousness together with its evolution. Without a rational strategy of research we will never understand the reality of our universe and why evolution gave man an instinct for rationality and for relying on causality.

The reality in science, however, tells a different story. The trend of irrationality is supported by an increasing number of abstract and poorly transparent physical-mathematical models. More and more irrational theories are presently proposed to prevent the irrational knowledge strategy from collapse. Everything in our environment changes in one direction, but all basic laws of nature have been designed to be invertible in time. Ideas of non-causality, of effects without causes, of non-locality, of objects being simultaneously in two places, and of an empty space which manipulates movements and time are already pretty much consolidated. Newer ones followed with the Big Bang scenario of the origin of the universe, of energy popping out of nothing, a gigantic inflation of empty space, or with time travel. Dark matter, dark energy, multiple worlds and an ever faster expansion of the universe with galaxies approaching light velocity are now confusing our intellect. Is nature really behaving so strangely and are essential fundamental mechanisms in nature irrational? Or is it, as I suspected, science which is insisting on irrational explanations, because it does not know better? If this were true, science is indeed on the wrong track.

When I was a boy, growing up in a small village in the mountains of Friuli, Italy, I learned early on that there was no future in cultivating this harsh, rocky land. No alternative existed to leaving the place and learning a modern profession. Owing to my interest in natural phenomena I had one early expectation and wish: I wanted to select a profession which would help me to better understand nature, so I became a natural scientist. But while science provided reasonable answers to many questions, it also generated profound confusion by insisting on irrational explanations for relevant fundamental phenomena. Is nature really irrational in such basic phenomena as causality, can time and space be manipulated, and can energy and particles really pop up from nothing? If this fundamental irrationality were really implemented in nature, why do we not see irrationality somewhere emerging among the fantastic technological achievements of living nature.

There, everything that has been discovered and investigated is finally understandable logically. No room for irrational mechanisms is left. Were they invented by physical science?

When irrationalities are considered to arise from a lack of information within the scientific theories, then the problem to look at is the concept of energy applied. The reason is that information needs energy. When information is missing, energy is missing or handled incorrectly. Theoretical models may be incomplete or wrong where energy matters are dealt with. The search for problems with our energy concept therefore became the crucial working hypothesis of my intellectual initiative.

In this book I start, therefore, with a short excursion into the history of scientific thinking on energy, discover a contradiction and propose a modification of the concept on how energy behaves. Energy should not only have the ability to perform work, but it should have the tendency, the interest to do so. This just seems to be a simple modification, but it means a paradigm change, a change of fundamental models of thinking. One is no longer dealing with a world in a state of equilibrium as has been understood up to now, where fundamental processes are entirely time invertible. Instead the world becomes fundamentally irreversible. Energy conversion is driving action and the advance of time. Does this really reflect reality? In fact, it will be shown that the very fundamental principle of least action, which is the basis of many natural laws, leads to exactly the same conclusion when mathematically re-examined.

When this idea is elaborated, it is realized that one irrationality and paradox of physics after another can be eliminated. A key element here is a necessary reinterpretation of the particle wave dualism in quantum physics. The dynamic property of energy causes matter to change from the particle form into the wave-form. The quality of energy in a small particle and a widely spread wave is, however, not identical. Here a fundamental mistake of traditional quantum theory is discovered. Energy in the wave is distributed and needs information to be reassembled back into a concentrated particle. It is this information on energy distribution in a modified quantum theory, I call it an "information self-image of matter", which further eliminates additional irrationalities. It can, surprisingly, also explain the observed always constant light velocity as a local phenomenon of light. There is no longer any need to postulate a four-dimensional space-time world, which also manipulates reality and controls light velocity. This entirely unexpected result eliminates further irrationalities. As a consequence, the cosmic redshift of starlight and the big bang explosion of the universe are re-examined and understood in a different way. An alternative model for the universe is proposed. The subjects studied also deal with evolution and the question how arbitrary mutations can, in the long term, lead to highly developed, intelligent life. The conclusion reached is that contrary to present understanding, evolution has an aim. It is driven by the tendency of self-organized systems to maximise energy turnover.

Nature is not irrational and can be explained by rational models. This adds credibility to the proposed dynamic understanding of energy and nature and opens a way towards a new frontier: understanding the mind and consciousness as a straightforward consequence of a dynamic energy. It allows self-organization of information, thereby generating a kind of "living information", of consciousness and spirit. It can even be shown that creation of spirit may be the real aim of evolution. Such a surprising result changes nearly everything we presently believe in relation to the meaning and function of the universe.

ORIGIN AND CHALLENGE OF IRRATIONALITY

1 Irrationality in physics and in society

I believe that evolution itself imposed rational thinking on man. Those who did not think or behave rationally had a higher chance of failure and of being eliminated. One important element has always been causality. Every effect has a cause. Here energy laws are obviously involved. Energy has to flow to generate changes. Causality laws simply reflect the behaviour of energy in relation to changes. The laws, which govern nature, simply appear to follow rational rules and man had to adjust to these. He observed the rules in nature and acted accordingly. Rational thinking is simply bio-analogue thinking. It follows the lawfulness of nature. Organisms in nature follow rational laws. Why should inorganic phenomena be an exception? When early philosophers started to explore mechanisms in nature more profoundly, they continued the route rationality offered towards a deeper understanding. From the different choices they selected rational ones and excluded irrational ones. The opinion expressed here is consistent with the conclusions of important philosophers. Immanuel Kant (1781), in his "Critique of Pure Reason" said that "all changes occur according to the law of the connection between effect and cause". Science philosopher Karl Popper (1997) confirms, that "the belief in causality is obviously genetic and a priori". It was given to us by evolution.

A perfect example of rationality and logics is mathematics, and physicists are convinced that all natural phenomena can be described by it. Let us look at a complex mathematical treatise with many rational computation steps, which finally lead to the expected result. If only one step were irrational, or mathematically wrong, the entire mathematical effort would fail. The result would be wrong. Nobody would doubt that. If physical reality can be described by mathematics, as physicists agree, why should it then be permitted to introduce irrational steps here, for example non-causality, or time travel, which contradict logics? Either the mathematical architecture of physics should run into problems, or the presence of irrational mechanisms within rational ones in physics. Permitting irrationality in science should consequently be a move along the wrong track. In this book I will try to demonstrate that this is indeed the case.
For many centuries scientists searched for logical explanations for processes occurring in nature. But then quantum theory and relativity theory significantly changed the situation at the beginning of the 20th century. Irrationality, supported by sophisticated mathematical models which few can really assess became acceptable. The relativity theory, mainly developed by Einstein, astonished with a time and with objects which could be manipulated, as well as with a four-dimensional space-time that appears to control both movements of

objects and time and is expected to shape our universe. People became exposed to a world which had not been projected into our brain by evolution. Quantum theory followed with irrationalities. It had started rationally with Max Planck and his successful interpolation of the radiation from a heated black body and with Einstein's explanation of electron emission from solid interfaces. Quantum theory was then continued by numerous scientists.

Among them were Sommerfeld, Born, Bohr, Schrödinger, de Broglie, Pauli, Heisenberg and Dirac, who all basically tried to fit a mathematical formalism to experimental quantum observations. The results were astonishing. Einstein himself did not feel he was able to follow them all the way. He could, for example, not accept giving up the causality principle: He wrote to Max Born, " the thought of an electron in a ray choosing 'at free will' the moment and direction for abandoning it is unbearable to me" (Daecke, 1991). But quantum theorists insisted that there is no cause for a quantum phenomenon such as, for example, the decay of an unstable particle or a radioactive atom. Only statistical predictions can be made. It is fundamentally impossible to get more information about such a process. But quantum theory offered still more irrational surprises. An object, for example an elementary particle, could be present at two locations simultaneously, and there could be a kind of instantaneous information transfer within distant members of a quantum system. Also, there is the statement that a particle would only be there when it is measured.

Quantum ideas gradually entered cosmology and surprised with more irrationalities. The number of galaxies estimated to exist in our visible universe is approaching 500 billions (Galaxies, 2013). One galaxy typically accommodates alone more than 100 billion stars. This unimaginable amount of matter and energy is all supposed to have poured out from an exploding Big Bang event starting from a seed smaller than an atom. Here space-time is expected to have started. This is not all. Even though a time definition under Big Bang conditions would seem to make no sense at all, astrophysicists present a chronology of the universe down to an incredibly short instance of 10^{-43} seconds (one divided by one with 43 zeros). Ten significant Big Bang epochs are distinguished until the universe became one second old. Steven Weinberg, an American Nobel laureate, even wrote a book on the origin of the universe with the title: "The first three minutes". But what time was it, if time, according to Einstein, is an illusion and clock time is strongly affected by gravitation? (Chronology of the Universe, 2014). Even more astonishing: this incredible energy dynamic is deduced from a general relativity theory which is derived from entirely time-invertible fundamental laws and does not even respect the energy conservation law.

Present models of the universe include multiple universes, allow energy popping out from nothing, and calculate its behaviour assuming an unidentified dark matter. Scientists also discuss the existence of an invisible energy of gigantic proportions, which is inflating the universe and pushing galaxies to speeds approaching light velocity.

Is nature indeed behaving in such a bizarre way or could it be that our civilization does not understand sufficiently and is misinterpreting observed phenomena? Did science get on the wrong track by tolerating and approving irrationalities and is it now inventing more and more sophisticated irrationalities to keep an uncertain, incomprehensible knowledge structure from collapsing?

I decided to find out and started reflecting about irrationality and what it may mean for science and society. Is it possible to trace the origin of irrationalities and to discover whether they are really fundamental, as science today is claiming?
What, in fact, is irrationality? It is defined as cognition, thinking, talking or acting without inclusion of rationality. It is an opinion or action given through inadequate use of reason, or through cognitive deficiency (Irrationality, 2014). This is, of course, not as science is expected to act nor as scientists want to act. But somehow, through theories which seem to give practical results in spite of irrational aspects, through complex mathematical models which do not any more reflect an understandable reality and because of complex historical and sociological reasons, it happens.
One century has passed since the rise of quantum and relativity theories, during which a large amount of research and exploration occurred. The irrationalities claimed for natural phenomena did not only remain in discussion, but they have definitively consolidated their presence in science and society. With the modern information technologies the awareness of irrationalities in nature has spread into most households and the discussed phenomena have started fascinating and wondering many non-scientists. The finding that there is something mysterious and to our brain not understandable around us draws special attention. It is something like confronting an esoteric experience. Has science become a partially esoteric initiative? Many people nowadays are attracted to esoteric movements. Uncertainty concepts seem to provide some kind of haven in an uncertain world or even a justification or explanation for its unavoidability.
It is interesting to note that a confrontation between irrationality in science and in society is not an entirely new phenomenon. It was already present during the early decades of the last century. Established quantum scientists such as Arnold Sommerfeld and Max Planck complained about irrationality in society in the form of beliefs in miracles, astrology or esoteric power (Meyenn, 1994). Others, like the famous mathematician David Hilbert, or the philosopher and diplomat Kurt Riezler, criticised a decay of science due to irrationality. It seems that already then many people have seen a link between irrationality in science and in society. Einstein himself felt compelled to comment that "there is a strange irony in the fact that many people believe, that the antirational tendencies of these days find a backing in the theory of relativity". This underlines a sociological phenomenon, which, I believe, really exists: when respected scientists propose and discuss irrational world models, then this irrational world becomes acceptable, fashionable and tolerable also to imaginative people who are not experts. For example, if time can be manipulated in science and if time travel is possible and

witnessed in science fiction movies, why should an ordinary person not play with such an idea in everyday life?

There may indeed be a strong relation between irrationality in science and in society. I first realized that when longer time ago I was asked to comment on the very successful book "A Brief History of Time" by Stephen Hawking (1991) for a German broadcasting station. It is a well-written and mostly competent book, but in it Hawking also proposes in greater detail a model for the universe, in which time is closed in a loop. His conclusion was that there is no beginning and no end for this universe. It could continue forever. His model violated the most basic empirical law of traditional science, the second law of thermodynamics. The latter is based on entirely verified observations of energy and states that isolated systems always act to increase disorder (entropy) by accumulating energy in small, chaotic quantities, which can no longer be practically used. But Hawking goes even further and uses his model to claim that no God is needed for such a universe. In principle, science has no authority to publish conclusions about the existence of God, and to derive such a conclusion from a violated, established natural law is irrational and additionally unfair. Nevertheless, the book made a tremendous impact on the public. The otherwise quite critical German journal "Die Zeit", for example, wrote in "Zeit-Magazin": "The physicist Stephen Hawking is on the point of finding the formula which explains the universe". The book became an exceptional bestseller and was translated into 40 languages.

Money can be well earned from fantastic and irrational science creations. This is particularly seen in science fiction movies, where "time travel", "beaming" of people and "wormholes" through the universe have become a motor of cinematic creativity. But the market value of irrationality does not leave even respected science institutions untouched. The European research facility CERN, for example, justified the 7 billion Euro particle accelerator with experiments probing Big Bang conditions. This, first of all, anticipates that the Big Bang explosion was a reality. Then it assumes that the tiny seed universe, which supposedly gave birth to the estimated 500 billion galaxies in our known universe, can be simulated with our technical efforts.

The feeling that nearly everything may be possible and is governed by obscure, incomprehensible laws may indeed have a significant impact on the mentality of people and their politicians. An example of the penetration of irrational ideas from physics into society may, for example, be seen in a sketch presented by a cabaret artist (Gery Seidl, 2014) in a cabaret in Graz, Austria. His presentation included the following statements: "I was eating a sausage, when an acquaintance, a physicist, passed by. The scientist said that exactly the same sausage may at this moment be eaten by another person in distant Salzburg. And the best is that the cow which has provided the sausage has not yet been slaughtered. It is still alive. But they (the physicists) do not care anyway. For them neither time nor space exists". This sketch was obviously an allusion to non-locality (the same object at two locations), Schrödinger's cat (which can be alive or dead), the time concept (time is an illusion) and instant quantum communication (space is not relevant). Via such widely popularized irrationalities, people do not get only a strange impression about the activities and accomplishments of scientists, students must also get confused and discouraged when considering an education in physics or engineering. In Austria such an irritation is especially understandable. Since 2007 a rather successful satirical scientific cabaret with the name "Science Busters" has been active in commenting science with humour. A theoretical physicist, an experimental physicist, and a cabaret artist are doing quite a good job of making fun of subjects such as string theory, inflation theory, multiple universes or black matter. I have witnessed TV presentations. The humour, I am afraid, is at the expense of science. I remember seeing a discussion on string-theory on television. The large number of studies and the substantial size of the active research community elaborating upon this theory were commented on. Then came the question whether there were convincing results, making sense of the numerous dimensions involved in the models. This was denied. But there was agreement that at least many scientists have got a job opportunity.

There is another example of how irrational science claims enter everyday life: "quantum healing". It is an esoteric practice with a wide offer in training, seminars and healing activities, aimed at health problems ranging from the burn-out syndrome to overweight. Quantum concepts such as Schrödinger`s cat, non-locality and non-causality are used to justify and explain its function, which, however, appears to have no scientific backing at all. Quantum paradoxes are apparently simply used as a selling strategy to suggest access to mysterious abilities.

I heard about gifted students who did not go on studying physics because they felt they could not deal with concepts they did not understand and will never understand. I also remember the irritation I was faced with when first confronted with quantum theory. Irrationalities communicated in science give some students the impression of not being able to deal with the challenges of the discipline. They turn to other professions. On the other hand it may also be that irrationalities in science attract students who are open to esoteric influences. Is this what science wants? Indeed, science fiction phantasies motivate young people to engage in technical challenges. But what is the contribution of invented phenomena such as time travel, wormholes, beaming of people and near light-velocity space travel to the spirit and to visions of modern man? Do we add building stones towards an irrational future, a future where logic reasoning is no longer a priority? Here one should also ask the question whether the interest in and the addiction to drugs in some population groups is an expression of increasing neglect of logic reasoning in human society. Why are young people so carefree about risking damage and manipulating their brain via hallucinating drugs? In a primitive society, where they would have been forced to face reality, they would not have survived easily. If applied, drugs were used only ceremonially and in a very restricted way.

Irrationalities, when presented as facts, lead to confusion. Facts are no longer taken as seriously as they should be. Laws, conventions and responsibilities of all kinds may be questioned. This concerns also the "well-educated", academic professions. How could it otherwise have happened that a young generation of well-trained, clever investment and finance people triggered an international bank crisis? The participants, busy extracting money from society without contributing to the real economy, must have known that banking institutions were sliding towards chaos.

And how can one otherwise understand why global environmental problems are handled so negligently by international politics. Why numerous environmental conferences are following each other without relevant decisions being made? Politicians know that our world is gradually deteriorating and gets little chance to recover. But there is nevertheless the expectation that things will eventually somehow work out well. I myself have witnessed the labour pains of a global environmental consciousness. In 1979, 35 years ago, I published a book with the title "Back to the Sun", in which I explained the need to return to a sustainable energy economy (Tributsch, 1979). Numerous other scientists have also warned of impending danger for the world climate. The arguments today are still the same. But rational arguments seem to be less attractive than irrational theories. The circumstance that some scientists are, for different reasons, still doubting the greenhouse theory is used as justification for doing nothing. And certain countries have adopted the comfortable politics of waiting for definite proof from science. And this will not be coming, because there are always some scientists who disagree. However, one should definitively act and stay on the safe side and should make provisions for the future, even if there are still some uncertainties, because this would be the best strategy to safeguard our globe. Such a strategy is based on experience of mankind confirmed again and again since ancient times. The wise Chinese philosopher Confucius, who lived around 500 BC, comments it with the following quotation: "Whether God exists or not we do not know. So let us make him sacrifices".

Politicians also know the devastating potential of nuclear and chemical weapons. They nevertheless engage in strategic adventures, as in former centuries. And they use an increasing amount of disinformation to manipulate people. The feeling here is there may be non-logic possibilities to avoid disaster and to handle the fate of mankind. If science lives with irrationalities, why are they not acceptable for real life? Will future generations judge our present society as one seduced by irrational concepts? Is it a society which is hiding real challenges behind a mirror of self-created irrationalities? During the last half century a series of international wars were not initiated for logical, publicly justifiable reasons, but on the basis of manipulated incidents and aimed misinformation. Even in educated democratic societies people overlook and tolerate invented and incorrect information too readily. In fact, science should act as a bulwark against lies and irrationalities. Can it do that? Unfortunately, questionable scientific arguments sometimes also support dubious commercial aims, because employed or paid scientists feel they have responsibility to the companies involved. Is this also a problem of science education in schools and universities?

Due to the strong population growth on earth, the limitation of fertile land and of productive oceans, due to both the challenge of energy and pollution, as well as the fast progress in technology, a science which looks far into the future is needed. In most of its activities science is rational and doing, in my opinion, a reasonably good job. I make this comment as a scientist criticising his own profession. But I also believe that irrationality spreading in fundamental science is undermining the determination to work hard towards solving real problems. Is science to be blamed for that? Or is nature fundamentally irrational? Then science, of course, has no alternative and is justified. I decided to find out by searching for rational alternatives to irrational theories in fundamental science. This is, I admit, quite an ambitious venture.
But it seems to be a meaningful experiment, an adjournment in our fast moving world of science and progress. If I succeed in challenging irrationality in science, then the aim of this book will have been reached. If not, the effort may be an interesting testimony of how a trained scientist, as I am, understands the natural world very differently from mainstream scientists.

2 Perception and non-perception of scientific phenomena

In order to distinguish between rationality and irrationality in understanding nature we should first try to understand what cognition and understanding means to our brain. Philosophers have already for a long time attempted to address this question. A good step forward was made by the imaginative philosopher Immanuel Kant. He was born in 1724 in Königsberg, Germany, now Kaliningrad, Russia. He led an extremely disciplined life there and became famous for his rational thinking. Kant also tried to find out what it means to understand our world. He distinguished between a so-called "noumenal" world (from the Greek word noumenon), a world with things as they really are which we cannot directly comprehend, and an altered, "phenomenal" world, which our mind creates when perceiving it. It is the view of the world in our head. It may be quite different from the real world. Kant considers the recognized objects as the consequence of a synthesis of a manifold of perceived information. Basically, the mind brings categories, termed "schemas" (meaning concept or chart) in connection with intuition (Kant, 1958). That way our comprehension of phenomena in our world is functioning.

The philosophic school, which Kant initiated, had many followers. A most prominent one became Ernst Cassirer. He was born in 1874 in Breslau, then Germany, today Wroclaw, Poland, and became a very productive, well-informed

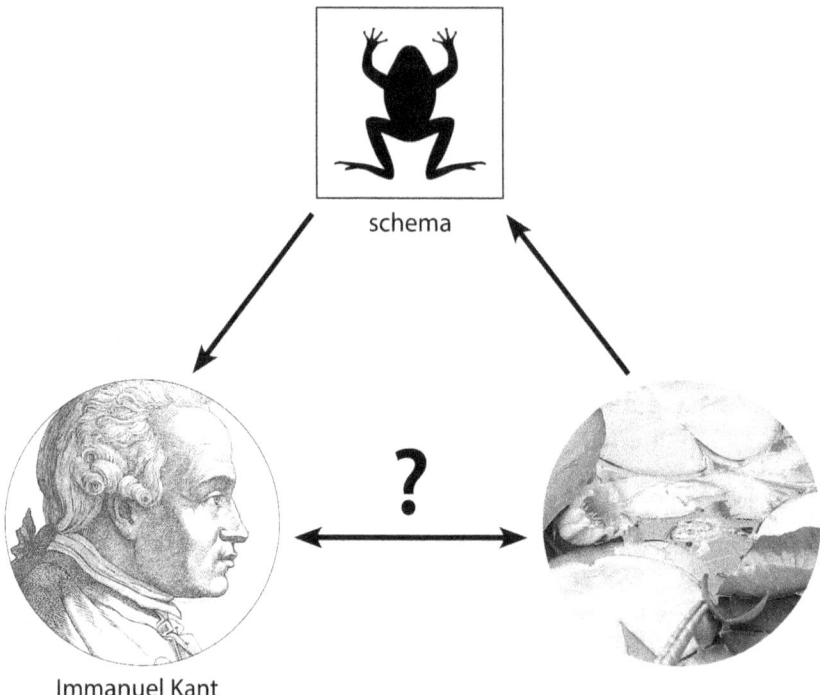

schema

Immanuel Kant

Fig.1 Immanuel Kant became famous for of his logical approach to essential questions in philosophy. He imagines a "schema", a synthesis of information and intuition, as an aid to help recognize the world as it is. Science philosopher Karl Popper (1979) understands this schema as an "intellectual aid" or "logical tool".

and widely interested philosophy professor. Owing to his Jewish origin and the tragic situation of Jews in Germany, he had quite a turbulent life, which forced him to emigrate to Sweden and then the USA. He published on many subjects interesting to philosophy. Among his many articles and books one finds "Kant's Life and Work" as well as "Einstein's Theory of Relativity" or "Determinism and Indeterminism in Modern Physics". For me, he is the philosopher who came closest to answering "what understanding means". For Ernst Cassirer (1923,1924,1929,1997) the human world of reality is "created" by "symbolic forms" of thought. His "symbolic forms" were essentially a reshaping and further development of Immanuel Kant`s "schemas". Cassirer defined symbolic forms independently for myth, religion, science, art and language. Each one regulates the world under its specific perspective. The "symbolic form" is the medium through which thought and cognition occurs. The symbols may have multiple connections with each other and are essentially a means by which a synthesis is created within the conceptual world. Each symbol is a factor within a system of symbols which are characteristic for different disciplines. The symbolic forms for art, for example, cannot help us perceive science or religion. For science symbolic forms include, besides statements in numbers and the law of causality, the most fundamental laws and principles of nature, shared and extended through discovery, understanding and communication (Cassirer, 1964; Schmitz-Regal, 2002). When keeping in mind these laws and principles, which are the relevant symbolic forms, and reflecting about a problem of science, one will rationally understand it. The helpful symbolic forms here are knowledge and the logic of science. They are oriented towards establishing convincing relationships between specific causes and effects and aim at finding universal rules of change.

Do we have a simple example of a symbolic form, which is helpful in understanding science? A natural symbolic form, which gives access to the understanding of living nature, may be DNA, the molecular code of information and heredity and its mechanism of operation (fig. 2). By knowing how it works, how information is passed from generation to generation, and how this information can shape individuals, and by knowing other natural laws, we can understand essential functions of living nature. By not knowing the function of DNA we could not reasonably well understand life.

Later, when talking about the somewhat limited information storage capacity of the human genome (chapter 29), it will be realized that an additional symbolic form, self-organization of information has to be added for a better rational understanding of the function of DNA.

Fig. 2. An important natural "symbolic form" towards the understanding of molecular biologic mechanisms and of evolution is the genetic code, the DNA. But visualizing the structure and its purely chemical function alone may not be sufficient to gain a full understanding

Cassirer made a clear distinction between symbolic forms for modern science and for primitive societies. The symbolic forms of the latter he understood to be controlled by myth, by failing distinction between individuals and surroundings, between part of an object and the entire object, by different concepts about causes and effects. However, here Cassirer underestimated the logic ability of primitive societies. He did not have experience with their way of life. Malinowski corrects him in this view (Malinowski, 1948). He demonstrates that primitive societies follow only partially a mythical mentality. As far as their real survival is concerned, their way of coping with natural phenomena, their strategies for hunting or for agricultural activities, they behave in a very logical and systematic way. They follow symbolic forms related to those which science is expected to take into account, but of course, on the particular level of their accumulated experience. Personal experience is also very important for perception. It is well known that education, experience and the learned profession are relevant for understanding and handling information. Depending on the background of people understanding and interpretation of recognized objects can be very different (fig. 3), but all will try to reach a logical understanding. Our subconscious mind already selects preferred objects when we, for example, go through the streets of a city. A mother remembers later only places where her children could play, a child a shop for candies, a motorcycle driver a repair shop for motorcycles. When analysed by persons with different educational background objects will also be seen differently. Personal symbolic forms are already active and help to find logic interpretations.

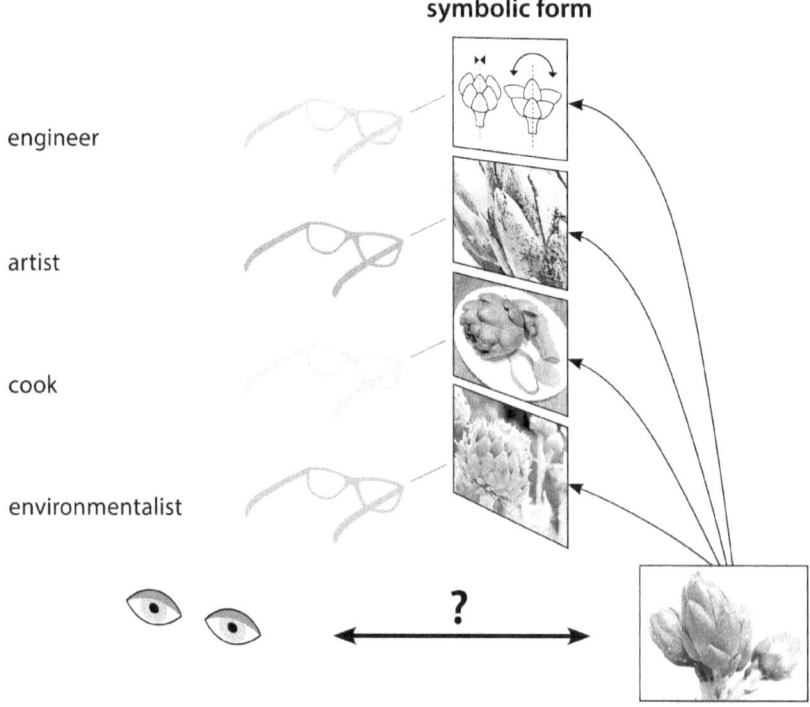

Fig. 3. An object such as a green plant, here an artichoke, is seen and understood differently in the eyes of people with different professions. It can be seen as an engineering object, as a piece of art, as food or as an impressive living system.

The finding that primitive societies rely on rational symbolic forms when dealing with survival, underlines my initially expressed statement that logic reasoning is a strategy, imposed by the laws of nature. It has helped man to survive for hundreds of thousands of years and it should be respected and cultivated because it may help man to also survive in the future. The biological role of logic recognizing as a form of natural adaptation is also evident for Karl Popper (1979). For the insight into scientific contexts he recommends: "if you want to recognize you must search for laws". Such laws will decide about the reliability of the "symbolic forms" applied. It will be the aim of this book to identify such laws which bring back logics into fundamental science.

How can this aim be accomplished?

The philosopher Cassirer did not realize that primitive societies behave logically when dealing with activities essential for survival, but he recognized that finding the basic laws related to reliable symbolic forms is a significant challenge for science. His vision of the nature of such laws has, for this reason, not remained constant during his long professional life. It was certainly also influenced by the quite dramatic increase of information on quantum physics and the theory of relativity, which occurred during his lifetime.

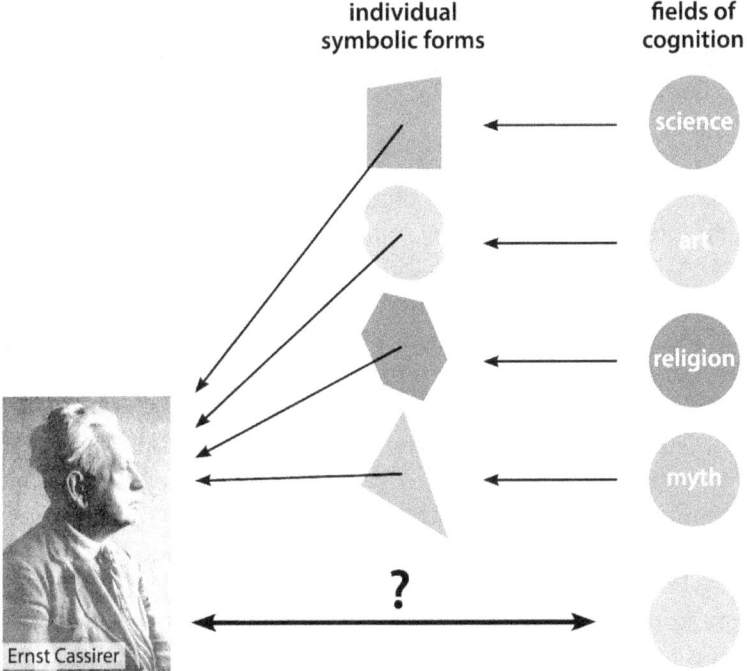

Fig. 4: Ernst Cassirer was fundamentally inspired by Immanuel Kant`s ideas and developed the concept of "symbolic forms" as tools for understanding. Selection of the respective appropriate symbolic tools mediates the understanding of science, of art, myth and religion.

On the basis of such considerations on symbolic forms we may reach the following conclusion: we understand nature and the scientific principles governing it only because in our attempt to understand, we recognise and involve unifying mechanisms. These are, for example, the principle of least action or the most basic laws on energy and matter and their behaviour in space and time. These and other fundamental principles and laws, which we use as a basic empirical and experimental network and the mathematical formalism which science has created around them, shape and determine what we call scientific perception. It is the process of attaining awareness or understanding of the environment via our senses and our brain.

However, more recent research in science philosophy has shown that the situation encountered in perceiving nature is probably even more complicated. It was basically claimed that a natural scientist applying the function and symbolic form of the DNA (fig. 2) and also the symbolic forms, the laws and principles of physics, may still not reach a consistent understanding of life. He will simply not understand consciousness and the human mind with its determination to act (Nagel, 2014) (for a criticism of standard theories, see also Saphiro, 1986). It is argued that in this case, the adopted fundamental principles, the "symbolic forms," may not be complete or may be incorrect, since they do not allow understanding of the function of the human mind. Natural laws must have permitted evolving it before man appeared on earth. I will take this argument seriously and come back to it later. First I will return to the problem of irrational theories in science.

What does the discussed concept of perception mean when scientists adopt a theory which is irrational, for example, that there is no cause for an effect? Does a failure of rational understanding in quantum physics or with the theory of relativity also mean a kind of scientific perception? Is it perception when established physicists like Richard Feynman, Roger Penrose or Steven Weinberg come to the conclusion that quantum theory, while matching experiments (because it was adapted to do it), does not make sense or that they never entirely understood it? Nobel physicist John Wheeler, for example, once ironically said "If you are not completely confused by quantum mechanics, you do not understand it". He apparently simply wanted to express that the way quantum mechanics works on a practical level and describes results is fundamentally incomprehensible to man. Where, then, remains our dream of a perception of nature through logic understanding when we are faced with the demand by A. Zeilinger, the well-known Austrian scientist working on quantum phenomena, that we should accept that a realistic description of the foundations of nature is not possible (Zeilinger, 1999, 2000)? He even commented: "we will well have to say goodbye to the naive realism, according to which the world itself exists, without our hand in the matter, and without our observation" (Naica-Loebell, 2001). This comes close to the well-known interpretation of the Kopenhagen school, which essentially shaped the present understanding or, better, the "non-understanding" of quantum phenomena. I say this because one has to ask the following question: What sense do symbolic forms make for the recognition of quantum phenomena when there is no hope that we may logically understand them? This really becomes a fundamental question.

According to Zeilinger (2007) an inevitable new ideological cornerstone is that an observer of nature does not merely disturb the observed system, but rather brings it about. Furthermore, local realism, the understanding that the observed details actually correspond to the properties of the system observed, has to be abandoned. Instantaneous action at arbitrarily large distances, as quantum physics claims it, contradicts local realism, which considers observed objects essentially as independent. Most important, the results of quantum observations become in addition accidental, without a hidden reason.

Here an unpleasant question is justified. If measurements only shape reality, if these measurements only bring about the observed results, why is society then spending so much money financing them, for example in the CERN facility? Why are we interested in experimental results which do not reflect objective reality of our world, but constitute doubtful information which the measurements themselves manipulate?

A. Zeilinger, who concentrates on experiments aiming at teleportation, recognizes still another new principle. While reality and information on this reality are separated in classical thinking, which means that reality does not bother about information on it, they are strongly interlinked in quantum understanding. "If information exists, it is expected to affect the outcome of measurements, regardless of whether one is aware of the information or not". Can one, on such a basis, hope to understand nature at all?

An unconventional, comfortable idea, dealing with the problem of non-understanding, has recently been advanced by the philosopher Richard Healey (2012). According to his pragmatic opinion "the real significance of quantum theory for the philosophy of science is how it advances the goals of physics without presenting us with novel ways of representing the world". This apparently means that we should accept the practical advantage of quantum theory as a working tool, without expecting clarification towards fundamental insights into physical reality. Is this really how we want scientific theories to function? Or is this a sign of resignation, a strategy allowing us to live with non-understandable, but mathematically describable phenomena? Healey argues that a physical theory may be completely successful without offering a representation of reality at all. I feel that if we have to accept such a pessimistic opinion, our world would become poorer, since there is no hope of understanding it even reasonably well.

Here in this book a different, a rational approach towards the meaning and interpretation of quantum theory, the theory of relativity, and theories about the universe is attempted starting with some of the philosophical ideas promoted by Ernst Cassirer.

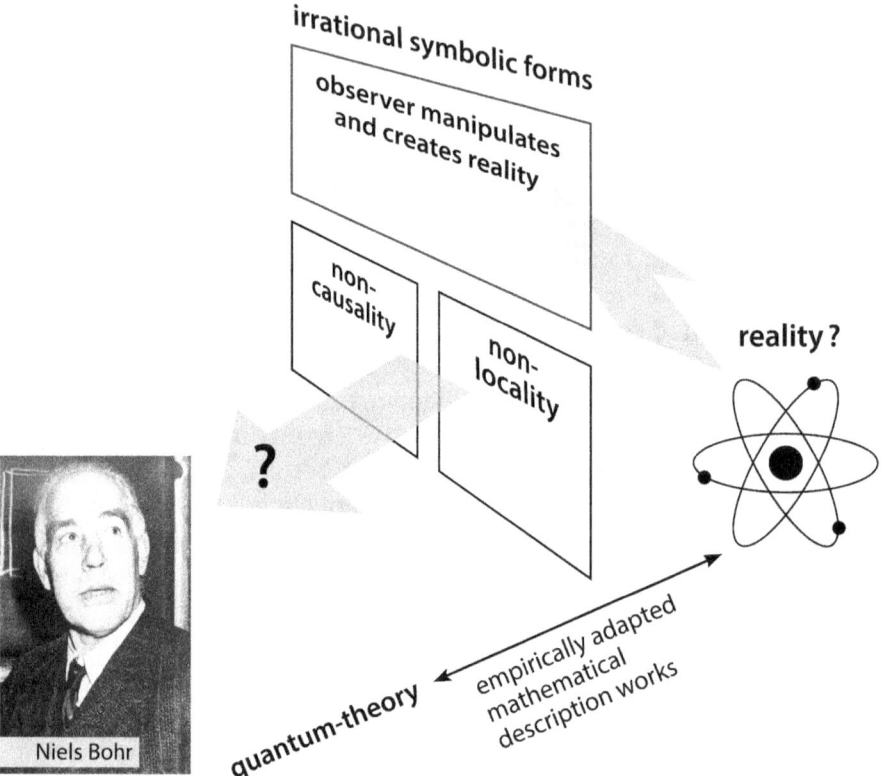

Fig. 5. Phenomena such as non-causality, non-locality and observer-manipulated reality make the quantum world irrational for an observer. Is this still rational perception? Was Niels Bohr right in assuming that the observer essentially manipulates and shapes nature?

In particular, it will also be tried to demonstrate that quantum theory only leads to bizarre paradoxes because essential information for the observer is suppressed or incorrect. This means an attempt will be made to restructure the "symbolic forms" applied for recognition of scientific subjects. Already here colleagues from quantum physics will fend off such a concept: in a well-known paper from 1935 Einstein, together with Podolsky and Rosen (1935), tried to demonstrate via discussion of a paradox that quantum mechanics is incomplete. Their conclusion was that either hidden variables should exist, which better define the quantum system, or information on quanta are not defined until measured. This nourished the hope that a more advanced theory could overcome or rationally explain the "spooky action" at a distance, which was predicted, a paradox which was criticised by Einstein. However in 1964 John Bell derived a theorem which states that no theory based on "local hidden variables" can ever generate the strange, verifiable predictions of quantum mechanics (Bell, 1964). The theorem aimed at a distinction between rational classical theories and irrational quantum models. A limit for measurement data is given, which realistic, classical theories have to respect. A non-realistic, non-local theory violates this limit. In such a theory, measured data are not independent of the measurement and of long-distance interactions. Quantum measurements were found to violate the limit which Bell`s theorem gives. Detailed experimental studies followed which seem to confirm that realistic, local theories can never explain quantum mechanics (Clauser et. al, 1978) (Aspect, 1981). On the basis of such a quantum understanding there is apparently no hope for a rational interpretation. In fact, Bell's conclusions have been confirmed and supported in numerous scientific meetings. However, as I understand these measurements, they do not prove that the quantum mechanical theory is complete, and they base the conclusions on an incorrect understanding of energy properties. It should also be mentioned that in 1930 the mathematician Kurt Gödel proved in a presentation at Königsberg (now Kaliningrad) that "only from a point of view outside a formal system can it be demonstrated that a system is logically consistent". Bell carried out the analysis from within the system. I will do it from outside with a new view and show that quantum physics can be logic and will also provide a logical explanation for the action at a distance within the phenomenon of quantum correlation.

But the reader should be patient and for the moment continue to follow philosophical interpretations of cognition of scientific subjects.

Ernst Cassirer represented the Neo-Kantianism of the Marburg school and was consequently searching for a rational and a-priori theory of cognition, a theory of knowledge, an epistemology, as philosophers say, that is governed by well-defined basic laws. Owing to the uncertainty relation, the theory of relativity, as well as the joint properties of space and time, physics no longer fulfilled the classical existence criteria of a substance (e.g. a particle with a given mass) in the sense of the philosophy of Immanuel Kant. As a consequence, Ernst Cassirer proposed to replace them with criteria for their function. These should be given a priori by natural law (Cassirer, 1937, 1921). Cassirer thought that this new approach, the inclusion of function into the "symbolic form" mechanism of perception, should ultimately lead to a deeper understanding of modern physics. Specifically, he, for example, thought that general covariance, the invariability of law against transformation into different reference systems, could be seen as a principle of objectivity for an advanced conception of physical reality. It is, in this case, no more the existence of particular entities propagating in space and time which reflect objectivity, but rather the invariance of relations during transformation between different reference systems (French, 2001; Ihnig, 1999; Ryckman, 1999; Itzkoff, 1997; Werkmeister, 1958). To achieve this invariance was actually the aim of Einstein's relativity theory. But does this theory with its irrational conclusions indeed reflect objectivity?

3 Perception with contradictory symbolic forms?

In order to become accustomed to dealing with symbolic forms, let us now develop an intellectual experiment (readers who are not familiar with mathematical symbols may just skip it and get the information from fig. 6). We will assume that we have two different concepts of fundamental laws, one correct and one incorrect, which we will use as symbolic form to generate scientific understanding in a wider area of physical phenomena. It may include quantum physics as well as classical physics. Let us choose very fundamental laws on energy (which consider that energy is equivalent to mass on the basis of relation $E = mc^2$). We will try to use these laws ((E_1) and (E_2)) of which only one (E_1) is correct as a basis for perceiving and understanding nature. Then perception P_1, (e.g. applicable to classical physics) which we achieve with the first set of energy laws (E_1) will be a function, a consequence of this set (f means: function of):

$$P_1 = f(E_1) \tag{1}$$

Alternatively, the perception we achieve on the basis of another set of energy laws (E_2) in an adjacent field of physical science (e.g. quantum physics) will be shaped by

$$P_2 = f(E_2) \tag{2}$$

If the difference between the two energy concepts E_1 and E_2 is significant, then also perception P_1 will be significantly different from perception P_2. If energy concepts E_1 (classical physics) and E_2 (quantum physics) are contradictory ($E_1 \neq E_2$), then also perception P_1 and P_2 should be contradictory and at least in part incompatible. If P_1 (classical physics) yielded an understandable science concept, P_2 (quantum physics), of course, would not. However, cognition could nevertheless be partially useful and satisfactory if the symbolic form E_2 were just partially incorrect. Essential mathematical mechanisms could still be functional, but a person searching cognition via the incorrect or incomplete symbolic form may be faced with rational contradictions.

Since information has an energy content (see later), the incomplete or incorrect energy concept E_2 could mean a lack of information for an observer searching scientific cognition. In other words: when contradictory symbolic forms E_1 and E_2 are used in adjacent fields of science, the scientific understanding of nature via P obtained with perception P_1 and P_2 cannot be joined into one general and unifying perception P including both fields (the symbol $\neq$ means unequal)

$$P \neq P_1 + P_2 = f(E_1, E_2) \tag{3}$$

It would be significantly contradictory or non-understandable because it would be based on contradicting assumptions of fundamental laws E_1 and E_2. Information for the observer could be missing. We would end up in a partially irrational situation of perceiving scientific phenomena, because perception cannot be based on contradictory symbolic forms.

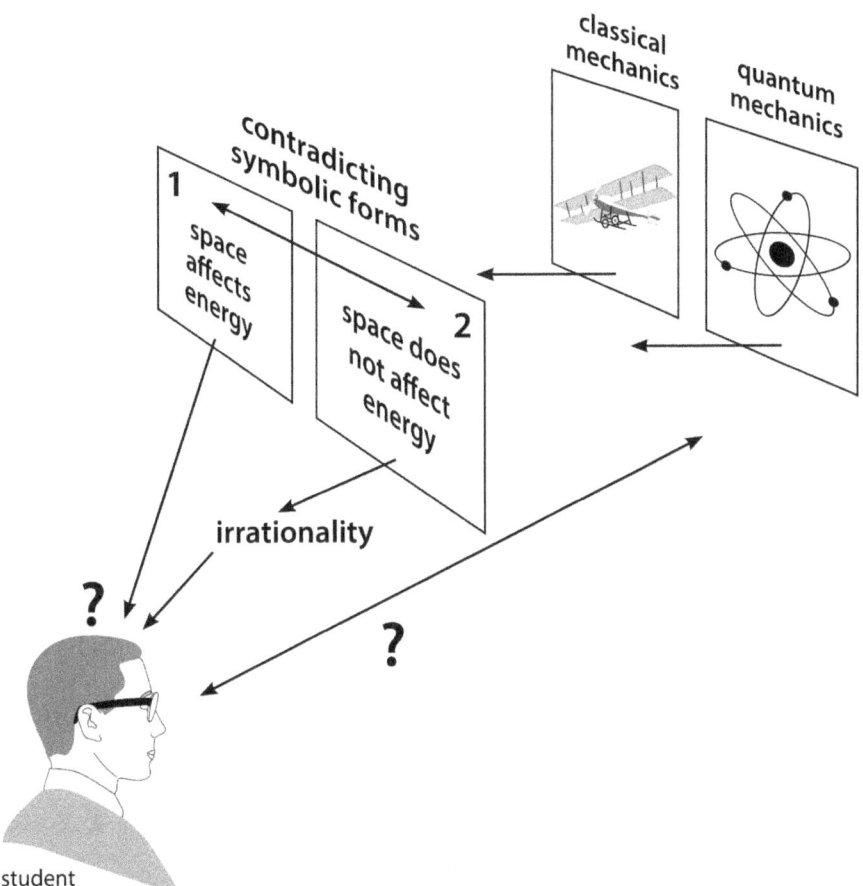

Fig. 6. If one tries to understand adjacent fields of science relying on non-identical energy concepts as symbolic forms, then a common unifying perception will not be possible. Much effort will also be wasted dealing with the irrational situation.

At this point it is useful to look at a kind of objective criterion for distinguishing between irrational and rational explanations of phenomena around us. I believe it is the information needed to make a phenomenon understandable which defines a difference. A logic explanation typically needs much less information compared with an irrational one. When one understands a situation, one stops asking further questions. An irrational explanation, on the other hand, keeps the questioner in a situation where he continues searching for answers. Quantum physics apparently does not provide sufficient information for explanations within the "symbolic form" structure, but forces the observer to waste a lot of additional energy on the turnover of information when racking his brain as to the irrationality faced (compare fig. 6).

As we will see later in more detail, information involves energy turnover. Energy has to be activated to create or turn over information. Man has apparently evolved to understand natural phenomena on the basis of minimum information (and thus energy) turnover. This has helped him to better survive. Irrational, esoteric, religious explanations are however tolerated when in relevant situations no rational explanations are available.

4 How to deal with irrationality

It is claimed here that the situation of non-understanding, described by (1) to (3) and fig. 6, does not represent a hypothetical model. It is the current situation in modern science, as far as quantum physics and the theory of relativity are concerned in relation to classical physics.

Quantum physics evolved during the first half of the 20th century as an empirical adaptation of mathematical relations to experimentally confirmed physical quantum results. The adapted formalism proved to be practically (mathematically) very successful. It became successful, because it was adapted to be so in terms of predicting numerical data, but it had not been optimised for yielding intuitively understandable concepts (as in classical mechanics). The formalism of quantum physics has, as we have seen, revolutionised not only science, but also philosophy with drastic scientific claims: the principle of causality (regularity causation) had to be abandoned (atomic nuclei and elemental particles are expected to decay without a given reason), and probabilistic causation invoked, where the relation between cause and effect in quantum physics is explained via quite bizarre probability assumptions.

The classical ideas on causation, as elaborated by David Hume (Hume, 1740, 1748) had to be extended and questioned and many philosophical studies and conflicts resulted from this situation (compare Hall, 2004; Collins et al, 2004; Baumgartner, 2014).

Another bizarre result from quantum mechanics is that individual objects are apparently simultaneously present in different places (double-slit experiment, non-locality). An observed quantity has, in addition, apparently no well-defined value before it is observed. According to relevant interpretations, the observer only appears to create it. So-called quantum correlated particles can, in addition, instantaneously communicate over arbitrarily long distances. The tunnelling phenomenon allows a particle to penetrate an energy barrier without surmounting the classically required activation energy and penetration occurs instantaneously without any loss of time. These are so-called quantum paradoxes

(Amelino-Camelia, 2000, Laloe, 2001, Aharonov, Rohrlich, 2005, Selleri, 1990, Berkovitz, 1998a, Berkovitz, 1998b, Dickson, 1998, Bub, 1979). They can typically not be perceived as logically understandable on the basis of practical experience and link such results of quantum physics with our understanding of "irrationality". How can something happen without reason, how can the same single object be simultaneously in two places, but when measured only in one? How can information be transmitted instantaneously over arbitrarily long distances? In addition, a most perturbing fact is the mentioned claim of quantum physicists (Bell theorem, (Bell, 1964)) of being able to prove that the irrationality of quantum physics is reality and cannot be overcome by expanding the theory, for example, via addition of hidden variables, meaning additional quantities that may better define the system. Under such circumstances, it is not surprising that, as already mentioned, quantum physicists claimed that irrationality and counter-intuition are fundamental characteristics of nature (Zeilinger, 1999, 2000). We should accept that nature cannot be understood rationally.

Attempts have, in consequence, been made to transfer such abstract, non-understandable, quantum aspects into the realms of cosmology, philosophy (Capra, 1979, Zukov, 1979) and speculations on consciousness (Penrose, 1994; Butterfield, 1995). As quantum physics is so successful in explaining properties of the sub-microscopic world, and because the physics community is convinced of having found the ultimate truth, we are gradually abandoning the determination, which has existed since the time of ancient philosophers, to find logical explanations for physical phenomena and to discard non-logical ones.

In fact, Greek philosophy distinguished between "true", rational assumptions, the domain of science and philosophy, and "false", irrational assumptions, the possible domain of spiritual activities such as art or poetry. We know today that the latter areas of human activity and many others like religion, which are based on sensual and emotional activities, can also be creative, following their own laws. Philosophers like Blaise Pascal and John Locke have amply emphasized this. But what about science? Is this abandonment of rationality in science justified and helpful? Will irrationality increasingly dominate our perception of nature? Will concepts of multiple worlds, of energy popping out from nothing, of non-causal events and dark matter increasingly substitute rational understanding?

As mentioned before, Richard Healey (Healey, 2012) suggests that we should accept quantum theory as a working tool for practically handling quantum physics without expecting a novel presentation of the world. We understand this proposal as the demand that we should not care too much about quantum paradoxes, but accept them as curiosities of a quite successful practical theoretical quantum tool. Should we follow such a strategy also with the theory of relativity and with cosmological models?

In contrast to such an opinion, in this book quantum paradoxes and irrationalities are being taken seriously. We are interpreting them in such a way that quantum theory (which was originally adapted to fit experimental results and not the understanding of observers) is still lacking some relevant information, which we, the observers, would need to really understand the proceeding phenomena. In this case we would have to think about the meaning of lacking information. What, in terms of physical quantities and parameters, does a lack of information mean in a theory which yields irrational paradoxes? Addressing this challenge it can be demonstrated that the answer is quite straightforward: it turns out that information has an energy content. For 1 bit the energy quantity of kTln2 has to be activated (k is the Boltzmann factor, T the absolute temperature, and ln2 the natural logarithm of 2). This is a very small, but not negligible, quantity of energy. At room temperature this amounts to approximately 0.7×10^{-21} calories (zero point seven divided by one with 21 zeros). Energy is thus needed to generate and turn over information. We can conclude: if a theory does not provide the necessary information for our logical understanding, it is not complete in terms of handling the concept of energy as we understand it from practical experience. Our working hypothesis must therefore be that there is something incomplete or wrong in handling the energy concept in quantum physics, as compared to classical physics. What could this be?

5 Two kinds of energy properties

The notion that the problem of irrationality in quantum physics is a problem of energy is an important insight. It can be verified by examining and comparing the energy concepts for classical physics and quantum physics. It may be that quantum theory works mathematically as its function of describing physical data does not require the entire information pattern of a complete quantum theory (this would support the suggestion of Healey (2012)). It merely uses an abstract aspect of quantum theory and does not need a complementary informational aspect, which human cognition requires.

This lack of information, this incorrectness in terms of energy, I conclude, may generate the known quantum paradoxes. If we wish to introduce the additional information into quantum theory, so that we can understand it logically and deepen our understanding of the physical meaning, we have to address the question of an incomplete or partially incorrect energy concept, in contrast to classical mechanics, which we intuitively understand.

Can such a challenge be resolved by looking at our concept of energy in a great deal more detail? Let us try!

I will demonstrate here that the irrationality problem of quantum physics is essentially based on the situation expressed by relation (3) in which irrationality is caused by contradictory symbolic forms expressing energy laws. We will consequently start trying to compare fundamental energy properties ((E_1) and (E_2)) in classical and quantum mechanics respectively (fig. 6).

Let us now compare the energy properties of some particles which are grouped together within a small volume and then expand into a large volume (fig. 7a). The mass of particles themselves is related to energy via the well known formula $E = mc^2$. It states that energy is equivalent to the mass multiplied by the square of light velocity. This is assumed to be an exact relation so that we are now looking at a picture of pure energy, spreading into space, both in classical physics (figure 7a) and in quantum physics (figure 7b). In this way, it does not matter whether we look at individual or many particles. When a group of classical particles (figure 6a) expands from a volume V_0 into a larger volume V, then, as the empirical 2nd law of thermodynamics says, less (free) energy will be available for work afterwards. The energy conservation law is, of course, not violated, but the entropy S, which is related to the disorder created and a measure for the no longer available useful energy of the system, did increase by

$$\Delta S = nR \ln(V_0 / V) \qquad (4)$$

(where R = gas constant, n = material quantity). The reader may be surprised that I present such a formula in a book which tries to explain complicated scientific subjects to a broader public. The reason is simple. It describes the formation of non-available, chaotic energy, when matter or energy is diluted. And this simple formula may, I claim, change the way quantum physics and our universe are understood (see chapter 22).

Entropy, which the above formula describes, is a quantity, which has been defined differently. Originally it was described as heat divided by its temperature. At low temperature the entropy value is thus much higher, in line with the notion that heat at low temperature cannot well be used for energy conversion. We may call such energy here "chaotic", or non-usable. According to Boltzmann large entropy is equivalent to large disorder or large diversity, small entropy to small disorder or high order. He identified the volume ratio in formula (4) with the probability of finding a particle or the information on this particle. The role of information can be deduced from a simple example: a bucket of Lego bricks can be identified with "large" entropy, a castle built of Lego stones with low entropy. The first contains no information, the latter much information.

It is seen that the entropy related to disorder increases with the logarithm of the ratio of volumes involved in the expansion. To calculate the contribution of non-available energy one has to multiply this entropy change with the temperature. This yields:

$$T\Delta S = nRT \ln(V_0 / V). \qquad (5)$$

The increase of "entropic" energy with an increasing volume is an experimentally verified mechanism and is valid for the expansion of electromagnetic radiation, as well as for an ideal gas. The entropy for expanding electromagnetic radiation (relation (4)) was first calculated by Wien in 1896 and then used by Einstein (1905) for the derivation of the Planck theorem. This entropy formula is consequently also true for light particles, photons, emitted from a star into free space, though with different quantities before the logarithm. It is an important fact that due to the very low temperature in space, the energy spent for the formation of entropy will be quite low. Energy losses by the photons will only be apparent at long distances. In any case the generation of entropy here will be an irreversible process. Simultaneously with the gas or light particle expansion our information on the system decreases. Since mass is equivalent to energy, it can be concluded that energy, when spreading and diluting into space, loses working ability and information. This confirms that space is a relevant factor for energy in classical physics. When expanding into space, energy properties change.

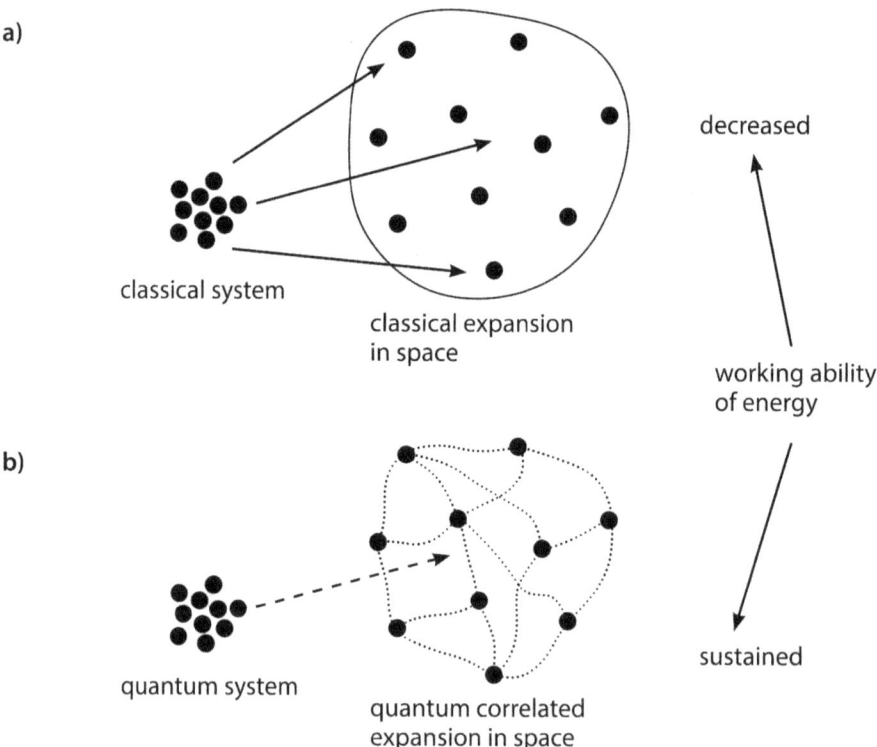

Fig. 7. Expansion of particles of mass m (energy $E = mc^2$) in space has different consequences for classical and "quantum correlated" systems. The working ability of energy decreases in the first (classical) case (a), not however in the second (quantum mechanical) case (b).

Let us now look at quantum physics. When the same particles are "quantum correlated" within a quantum mechanical multi-particle system (figure 7b), then the situation will clearly be different. To be quantum correlated means that two or more particles are linked to each other via a non-local interaction, which is peculiar to quantum systems. When specific quantities are measured the concerned particles share the properties to be measured. Joint properties appear at the expense of the loss of individual properties. Spontaneous interaction, correlation, between them will consequently occur. A change in one particle may have an instantaneous consequence in a distant, correlated particle. Large associations of correlated particles have in the meanwhile been observed (Krenn et al., 2014).

While the energy conservation law (corresponding to the 1st law of thermodynamics) and related energy concepts are in agreement with those used in classical mechanics, the consequences of energy dilution in space (equivalent to the 2nd law of thermodynamics) are obviously not. When energy is diluted, part of the energy should thereby be converted into non-available form (while the energy balance is, of course, respected). This is not the case in quantum physics (fig. 7b). Regardless of the distance over which the particles expand into space, quantum correlation and the operation properties of the quantum correlated system are maintained. Quantum theorists invoke distances across galaxies or even to other universes, distances over which quantum correlated systems should still fully function (and maintain their statistically calculated original entropy value). This is simply judged on the basis of the empirically obtained physical-mathematical formalism, which describes the correlated particles without considering the space between them. When, for example, the spin, which is an intrinsic form of an angular momentum, of one particle is inverted, another "quantum correlated" particle will respond instantaneously with a correlated spin inversion at an arbitrarily large distance. This is the quantum phenomenon, which Einstein named "spooky action at a distance". "Quantum correlation" here means there is a joint expectation value of the product of the outcome on the two sides. For specialists: in Bra-Ket notation the superposed, quantum correlated state of two particles A and B in the singlet state can, for example, be described by

$$|\psi\rangle_{AB} = \frac{1}{\sqrt{2}}(|0\rangle_A |1\rangle_B - |1\rangle_A |0\rangle_B) \qquad (6)$$

For this correlated state, due to the pairing (or, in the case of many particles, grouping) of particles it is possible to statistically calculate a reduced entropy. But whatever the space reached between the particles when they are expanding, the quantum correlated particles will always behave subject to the same mathematical formula. This formula does not contain a space dependent variable (relation (6)) and describes a simple two-particle system (multi-particle expressions with N particles are correspondingly more complicated). It also maintains its initial entropy value and remains subject to identical energetic conditions in spite of its expansion into space. This is obviously a significant difference in the understanding of energy between classical physics and quantum physics, which was searched for.

There is also a problem with information (which is related to energy) in quantum physics, which leads to a similar conclusion. When one spin inversion causes a quantum correlated opposite spin inversion in a quantum correlated system at an arbitrarily far distance, we are dealing with a transfer of (internal) information (which quantum physics in fact tries to exploit (Ma et.al, 2012)). Since information has an energy content, transfer of information is a transfer and exchange of energy. General classical physical experience is that the further energy has to be transmitted, the greater the parallel energy losses. Only long range information transfer, or (equivalent) energy transfer, in quantum correlated systems seems to be an exception. It is expected to work without energy input over arbitrarily long distances, which is quite strange.

Energy in quantum physics is consequently used as if space had been eliminated. Space simply does not exist. Distance does not affect the state, availability and transfer of energy in correlated quantum systems.

It is interesting to note here that in the theory of general relativity, which is, by the way, not compatible with quantum theory, energy in the form of matter behaves very differently. It even exerts an active influence on space. It can curve and manipulate space and even time, an incredible property which will be commented upon later. This may be one of the reasons, why quantum physics and the theory of relativity are not compatible.

Energy does, for a quantum correlated particle system, not decrease its ability to perform work when expanding into space. This is different for classical physics where expansion into space generates entropy, which is related to the resulting non available or non usable energy generated. In quantum physics space has simply lost its relevance for energy. This property was apparently overlooked when quantum formulas were empirically adapted to fit experimental results. Energy stays the same in spite of its dilution. There is apparently no mechanism in the quantum formalism, which changes the quality of energy when this energy is diluted. This identified different behaviour of energy in quantum physics, as compared to classical physics, is expected to be the origin of paradoxes and irrationalities in this specific discipline. We have, therefore, to learn more about the relation between energy and space in order to understand the basis of this mechanism. We should also try to find out why quantum physics did not consider it and to learn where the effect of space on energy could best be introduced into quantum physics.

6 Energy dilution in space and consequences

It is argued here that space must be a relevant factor for energy for a quantum system too, at least for energy`s ability to perform work. Energy itself remains, of course, a so-called extensive property of a system. A larger portion of a system, or a system with more matter will contain more energy. What is postulated here is only that a system with a given energy - when expanding into free space, or in other words diluting itself - loses some ability to do work. Not the total energy quantity itself, but a "dynamic" property of energy, its ability to do useful work, changes during its expansion into space. For a dynamic energy theory, which this book will shape, a dynamic property of energy is, of course, important. During the expansion of energy some available energy is converted into non-available energy. Entropic energy, non-available, chaotic energy, is created.

The field of reversible thermodynamics, which is based on practical experience and forms the foundation for understanding a wide range of energy conversion processes, gives clear examples of energy`s property of decreasing its working ability with decreasing energy density in an expanding occupied space. A vapour machine typically works between a heat bath at higher temperature and the temperature of the environment. It is known from the so-called Carnot law that a heat bath of a lower temperature has a lower energetic value and an inferior working ability during thermal energy conversion processes compared with a heat bath at a higher temperature. The same heat energy at lower temperature means one needs a larger heat bath if one would like to dispose of the same quantity of energy. But even with the larger heat bath the output of energy conversion will be inferior because the energy conversion efficiency will be lower. The difference in performance is described by the concept of entropy. Entropy is a measure for the disorder of a system. Entropy relates also to the number of microscopic states in which a system can be. It is also a measure of our ignorance about a system. High entropy means high ignorance and an energetically "chaotic" situation. Molecular movement in the heat bath at lower temperatures reflects less kinetic mechanical energy than in a heat bath at higher temperature. Similarly a gas space at high pressure is able to perform more mechanical work than a gas space at very low pressure. Again, a higher concentration of mechanical kinetic energy has a higher working ability.

Let us now find an example outside thermodynamics and compare the working ability of concentrated photons and highly diluted photons. A leaf of a green plant grows in sunlight with a quantum flux density from solar light of approximately 10^{-3} Einstein/m^2 (1 Einstein is 6.10^{23} photons/sec - 10^{23} is a one with 23 zeros). Under similar conditions a solar cell may convert quantum energy into electricity with 10-15% efficiency. In moonlight, with a quantum flux density of the order of 10^{-7} Einstein/m^2, a solar cell will already be useless and a leaf will not grow. When the photon density is still much lower, photons can still induce physical-chemical stimuli, but special detection systems are required, which even have to supply external energy for detection, as is done by a retina of an eye adapted to the dark or an electrically powered photon multiplier. Even in photo-electrochemical systems, for example in light-induced water splitting in an electrolysis cell, a threshold photon flux is needed to generate the chemical potential difference required for hydrogen and oxygen formation (Gregg, Nozik,1993). If this threshold is not reached, water splitting, and thus energy conversion, does not occur. This simply means that an electromagnetic photon field of high density is more valuable and more efficient for energy conversion than a low density field. This is, of course, well known also from other examples. Sunlight does not ignite a forest, but a lens, which concentrates the light, can (fig. 8). We will later see that the difference makes information in the form of the lens.

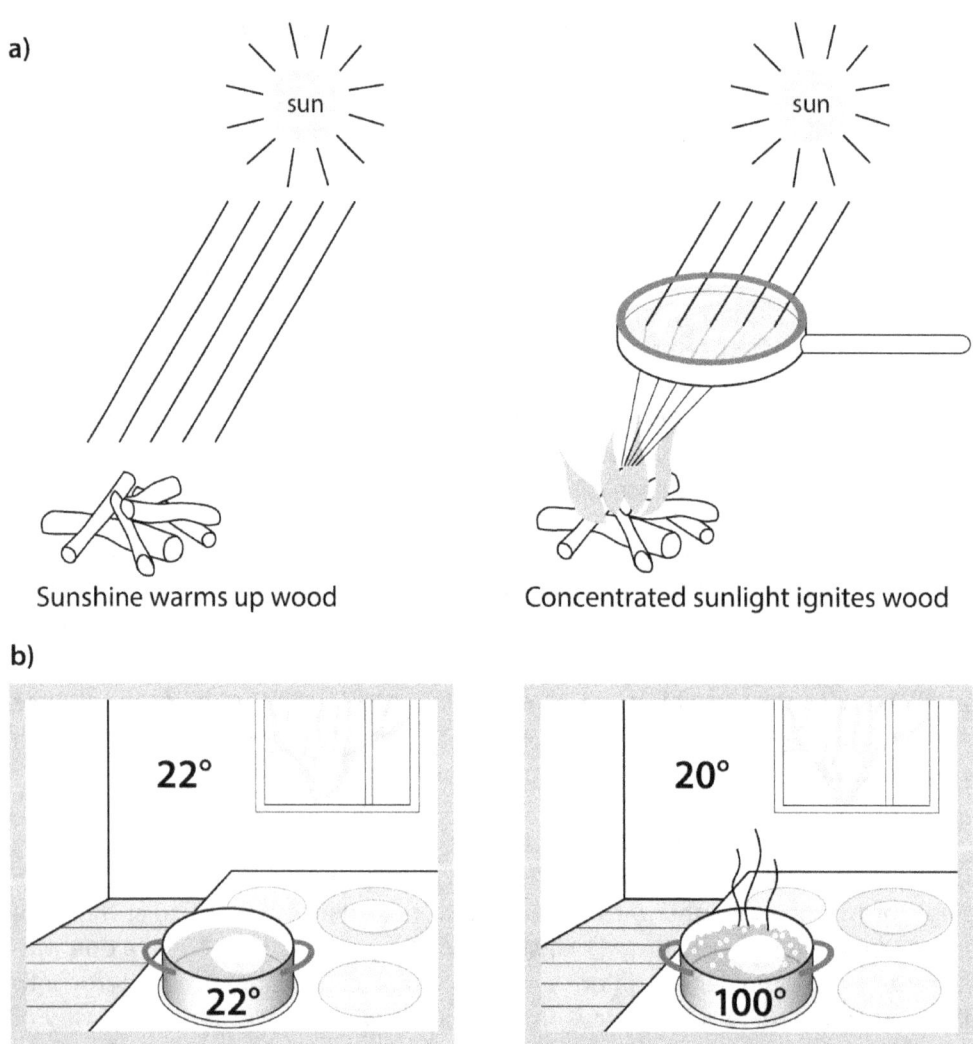

Fig. 8. Energy is more useful (or dangerous) when concentrated. a) The example shows how fire can be ignited using a lens in sunshine, but not by sunshine alone. b) Sufficient heat on a hot plate can get an egg boiled, but the same heat distributed in a kitchen cannot.

There are also practical consequences for energy harvesting. No imaginable machine could derive a similarly large fraction of energy from a highly diluted (expanded) energy field as from a concentrated (contracted) one.

Similar considerations can be developed for energy conversion via electrical, magnetic or gravitation fields. With decreasing density the energy`s ability to do work will increasingly become smaller and more inefficient.

The behaviour of electromagnetic radiation is described by the well known Maxwell equations, named after a famous Scottish scientist from the 19th century. They clearly formulate a supposed time-invertible mechanism - one whereby radiation can develop equally into positive and negative time direction. If electromagnetic fields of higher density had similar working abilities like fields with lower density, they could reversibly be expanded and contracted as Maxwell`s equation in principle allow. However, the reality is that the concentrated radiation field around an antenna expands only towards the horizon. But radiation from the horizon will not, without additional energy from outside, concentrate towards the antenna. Radiation processes are irreversible. This was already clarified in a polemic discussion between Planck and Boltzmann in 1897 (Planck`s Radiation theory, 1897). A phenomenon of a spontaneous concentration of radiation energy from the environment towards an antenna has never been observed (though some people still propagate it under the name of a "Tesla free energy generator" in the hope of an inexhaustible energy source from the environment and space).

Are there additional indications for a radiation which, in contrast to the present time invertible theory, follows a time orientation when disseminating into space? Interestingly, even our cell phones could provide evidence. In contrast to early generations of cell phones, today's no longer show a visible antenna, which before even had to be pulled out manually. They have an antenna, but it is integrated into the body of the cell phone and is very small. It nevertheless performs and functions much better in a much wider frequency range. The secret is its fractal geometry with self-similar structural properties.

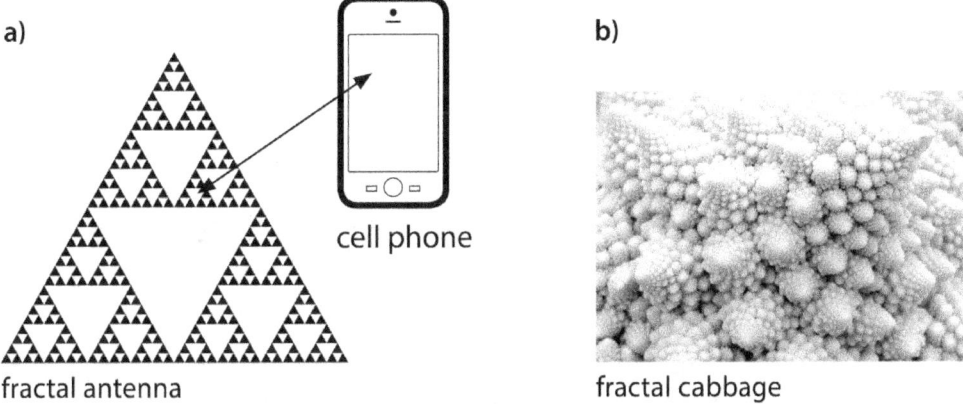

a)

b)

cell phone

fractal antenna

fractal cabbage

Fig. 9. a) Fractal, self-similar antenna, give modern cell phones a high level of reception and wide band sensitivity. b) Interestingly, such fractal structures are widely applied in self-organising biological structures, such as, for example, in a cabbage. This points to a time-oriented, irreversible nature of both mechanisms.

In such fractal structures the shapes repeat themselves over several scale sizes and the shape always looks essentially the same at every dimension. Duplicating a mathematical operation again and again using a result repeatedly to start a new cycle leads to such fractals. And such a fractal antenna can also be used efficiently to send out radiation into space. This discovery was more or less a discovery made by trial and error (fractal antenna, 2010) (Rusu et al., 2010) and it became very useful for modern information technology. The special aspect of this is that self-organizing biological systems also benefit greatly from fractal geometries and fractal processes. The heart rhythm follows a fractal, self-similar function, as does the nerve system, the lungs, or the DNA sequence. The beautiful fractal shape of cabbage and of many other biological structures are well known (fig. 9b). In all cases, the generation and structuring of such patterns require feedback processes which rely on an oriented time, a before and after. When cell phones function much better with fractal antenna, then they are demonstrating fundamental similarity to biological systems in dealing with self-similarity. The reason for this similarity must be time orientation, because fractal, and related deterministic, chaotic structures, arise and function only when there is a direction and flow of time.

A diluted radiation field has thus a property which is different from that of a more concentrated radiation field. It obviously has a lower ability to perform work because expansion of radiation generates entropy, which is basically non-available energy. It has been mentioned that this was, in principle, already found out by the German Nobel laureate, Wilhelm Wien, more than one century ago with his "distribution law", which basically states that the maximum wavelength of a black body shifts to a shorter wavelength with increasing temperature. Everyone knows the described phenomenon. It accounts for the changing colour of a piece of iron when gradually heated. It starts from dark red, which has a longer wavelength of light and gradually becomes white, when the wavelength of light becomes shorter at higher temperature. This temperature dependent energy distribution of radiation turns out to be described by basically the same formula as a volume of gas particles subject to temperature and volume changes. In 1905, this fact was used by Einstein to justify his claim that light can also be understood as particles, later called photons (Einstein, 1905). When a radiation field expands, its energy distribution should consequently change as it does in a gas expanding into space. Entropy is generated, which corresponds to a situation of higher probability and lower information content.

As in the thermodynamic example of an expanding gas, information (which involves energy) from outside, provided by an imagined so-called Maxwell demon, is needed to return to a higher energy density of useful energy again. Since the ideas linked with this intellectual experiment involving a Maxwell demon are important for our future considerations, it is helpful here to learn more about it. During the development of theories and technologies around vapour machines in the second half of the 19th century there was also discussion about the construction of a machine that could draw power just from cooling down a heat reservoir with one temperature. This would be a ship that could sail just by extracting the heat and energy from the ocean. Such a machine is called a "perpetuum mobile of the second kind" and was considered impossible by most scientists. In 1871, however, J.C. Maxwell came up with an intellectual experiment to challenge this belief. He wanted to show how energy could be extracted from one heat bath at one temperature via a skilled hypothetical demon. He separated the heat tank into two halves and imagined that the demon kept an eye on the velocity of water molecules. Then the demon rapidly opened a slit between the two heat tanks in such a way that only molecules at high velocity were able to pass from one section into the other (fig. 10). This way, just through the action of the demon, a temperature difference, and thus a working ability of the system could be developed.

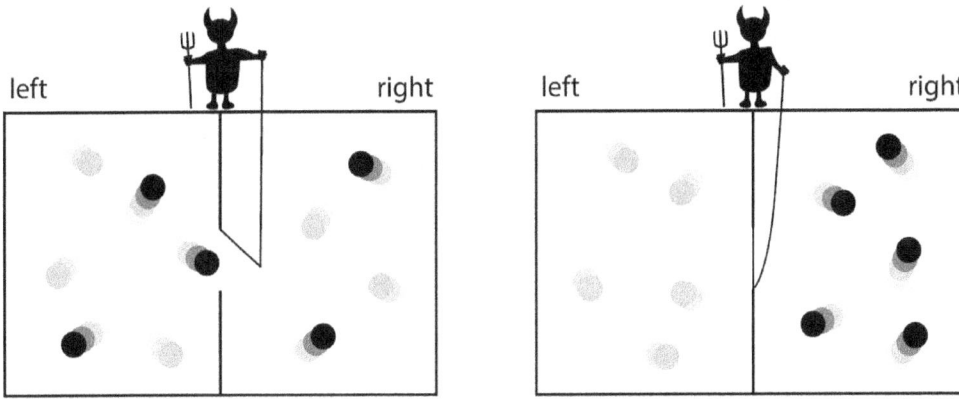

Fig. 10. By opening and closing a slit just in time, an imagined "Maxwell" demon can allow fast molecules to move from one heat compartment (left) to another (right), thereby increasing the energy content of the latter. In this way a machine could be powered, a ship could sail while cooling down the ocean. The demon has, however, to use information, which requires energy. And this information has to come from outside the system.

The situation created by the demon was puzzling at that time and nobody could in fact imitate his activity. The reason was discovered much later, via Szilard`s, Brillion`s and von Neumann`s contributions to the problem (Szilard, 1929) (Brillion, 1951) (von Neumann, 1966). It is that information, that which the demon needs to do his job, has an energy content. This energy amounts, as we have already mentioned before, to $k T \ln 2 = 0.69 \, kT$ for 1 bit of logic operation (k = Boltzmann constant, T = absolute temperature, $\ln 2$ = natural logarithm of 2). The demon has consequently to be supplied with energy from outside, otherwise he does not function. Only this supplied energy makes the recovery of entropic energy possible and thus keeps the system working.
An external supply of energy in the form of information to the demon converts

the originally impossible "perpetuum mobile" of the second kind into a permissible, functioning energy system.

In the case of spreading radiation which does not re-concentrate on an antenna, properly arranged reflecting mirrors on the horizon could change the situation. They could re-direct diluted radiation back to the antenna. The adequate arrangement of reflecting mirrors, however, requires information (provided by a Maxwell-type demon) and thus energy. No contradiction therefore arises concerning our statement that dilution of energy leads to a loss of working ability or of information. This example of electromagnetic radiation suggests that not only thermal systems but all energy systems are subject to such a deterioration during expansion into space. Let us look also at sound propagation around a microphone. In spite of an assumed time invertible fundamental physics, it has never been observed that sound accumulates spontaneously, which would also increase its ability to perform work.

Energy, which decreases its working ability when diluting into space, also responds to time. When acting that way it chooses the time direction which we experience. Consequently, there must also be a fundamental property of energy, which is not time-invertible, but oriented in time. This is not what present physics tells us, and has to be examined in greater detail.

DYNAMIC ENERGY: PARADIGM CHANGE TOWARDS RATIONALITY

7 A dynamic energy concept: why?

Up to now a possible lack of information in quantum theory, and the neglect of space for energy have indicated that there is something wrong with the presently applied energy formalism. I started to think more about energy and found a few additional arguments, which led me to the conclusion that it is indeed the present established theoretical energy concept which is at the origin of fundamental irrationalities in quantum physics, the theory of relativity and theories of the universe.

While, mostly on the basis of systematic experiments, the concept of energy has become a very useful and successful tool for modern science and technology, its theoretical foundation, in my opinion, remains in part questionable. The following four observations changed my understanding of energy. They convinced me that energy has to involve active change:

Consideration 1:
One of my thoughts concerned the historic evolution of the energy concept (as I discussed in Tributsch, 2008). Since antiquity philosophers and scientists have been looking at an ever changing environment and speculating about something which is conserved within all these changes. Much later energy finally turned out to be this quantity. It is not lost, but conserved, even though changed in its availability. Already the ancient philosopher Heraklitus of Ephesus, born around 540 BC, came intuitively close to such an understanding. Little has been handed down to us about his thinking, but we do know that he insisted on rationality, in the Greek language "logos", in recognizing and describing the world order. He was also impressed by the fact that everything around us changes. From philosopher Aristotle, who lived from 384-322 BC, we know about his comparison of "changes" with a river: "You never step twice into the same river". But he apparently considered fire (energy?), a supposed substance from the universe, as the source of all changes and as something stable: "The world is an eternally living fire". This fire could be changed into phenomena of our experience. He compared the generated driving force with a tension, such as released from a stretched bow or an activated string of a lyre.

The search for something, which is stable and sustained within all the changes around has always been a fundamental and still not completely answered challenge for mankind. During the same 6th century BC another personality was searching, in a spiritual field, for something stable and unalterable within the flow of human existence determined by birth, youth, suffering, age, illness and death. It was Siddhartha Gautama, otherwise known as Buddha. But let us return to energy, which is sustained during all occurring physical changes.

Fig. 11: Heraklitus of Ephesus reflected on the on-going changes around him and especially on what goes on with a river. And he speculated about the supposed stability of fire, a cosmic substance from which all changes were expected to arise.

Aristotle coined the expression "energeia", which was much later, around 1700, adapted by Johann Bernoulli and resulted in the modern term for energy. The development went on via the philosopher Leibnitz, who talked of a "vis viva", a "living force" around 1695, and Lagrange, around 1788. The latter added to the "living force" an energy quantity depending on the position of objects. His finding was that the sum became constant in time and energy was conserved. Many experiments on heat as energy were to follow and creative scientist such as Julius Robert Mayer (1814-1878) and James Prescott Joule (1818-1889) were needed to coin the modern concept of energy: the fire of Heraclitus became our energy, a non-destroyable, but transformable, substanceless, weightless property of dynamically changing systems.

The strange situation now is that our present concept of energy has lost any relation to the change around us from which it has originally been derived with so much effort towards understanding. Changes and dynamic behaviour have simply disappeared from the concept of energy. Energy is now just defined to be a number, a so called "skalar", a non-directed quantity. It has the potential to do work, but has no intention at all of doing it. One needs additional mechanisms to activate energy for work.

This strange property, by the way, did not escape the attention of Albert Einstein. He realized that energy is "asleep" and made an interesting comparison: " Energy behaves like a beggar who, in fact, is a millionaire. But nobody realizes it". This beggar does not spend and showoff his money like typical rich people. His life is accordingly very special and not characteristic of his wealth. Such beggar personalities existed sometimes, maybe because they were tight-fisted, or because they happened to be in special circumstances. I remember the extraordinary French woman Alexandra David-Néel, who in 1921, disguised as a beggar and accompanied by a red-cap lama monk, reached the forbidden Tibetan capital of Lhasa. In her book (David-Néel, 1994) she described the miserable life and the hunger she suffered as a beggar, while she had her travel-gold hidden on her body. She could not spend it, however, because she would immediately have been revealed as an intruding foreigner.

Later it will be explained that money is, in fact, the financial equivalent of the physical value of energy. One can use it to buy services, work and materials. And it will be proposed in this book that the laws of energy (money) should be changed from that of the stubborn, tight-fisted beggar to that of an ordinary money-spending person.

How did the energy concept in present-day physics lose its relation to change, from where it actually started? The Italian-French mathematician J.-L. Lagrange, around 1788, when studying his famous energy equations for dynamic systems, still considered and investigated conditions which reflected irreversibility and time orientation. This means that he paid attention to change. Also the Irish mathematician W.R. Hamilton, when, during the first half of the 19th century, he derived the now famous Hamilton functions, still argued that external irreversibility should be considered. He also felt that there had to be a relation to change.

Later the external conditions, which considered change, were simply abandoned from the concept of energy, probably during a search for simplicity and mathematical beauty. The constant "fire" of Heraclitus lost any dynamic relation to changes and to the decaying wood in the fire. Energy became just a number, without any relation to change. I myself have long reflected over this strange evolution of energy concepts in physics (Tributsch, 2008) and brooded over the consequences of a concept of energy, which maintains its relation to change. My conclusion was that we can only adequately explain changes in our environment

a)

"sleeping" energy / beggar

b)

"dynamic" energy / dynamic man

Fig. 12. In classical physics energy has, as Einstein observed, the properties of a beggar who is secretly a millionaire, but does not show it (a). Real energy, it is claimed here, rather behaves like an ordinary person (b). He attempts to conduct a decent life with his money.

and the evolution of nature if we allow energy to have a relation to change. Today all basic laws in physics are, however, formulated in a way that allows them to be time invertible. And energy is neutral towards time and changes.

This point is very important and my main effort will be focussed on linking energy again with change around us. I will try to take up the research track of ancient science philosophers. Only in this way will it really be possible to understand dynamic nature as it is with all its changes, and to eliminate present irrationalities and paradoxes in science models. The irreversibility of our world plays, as everyone can observe, a fundamentally constructive role in generating structures, evolution, life and decay. It has therefore to be implemented on a fundamental level. Reversibility is an illusion, even on a sub-microscopic scale. Scientists are not able to invert their experiments in time, even in well-controlled, simple experiments with elementary particles.

Consideration 2:
The second consideration, suggesting a dynamic property of energy, is related to the notions of quantum physics as described above, which neglect the relevance of space for energy. Today the duality of particle and wave in quantum physics is a generally accepted principle, but according to my considerations it is actually not a duality in terms of energy. There is a difference in the ability to perform work with concentrated energy compared with diluted energy. When energy in a particle and in a related wave is not the same in quality, there has to be a drive and an inborn mechanism, which changes the energy from that of a particle to that of a wave. Otherwise the experimentally established duality between particle and wave would not work. For this reason too energy, at least the energy which is able to perform work, has to be an oriented, a "vectorial" quantity. In its tendency to decrease its presence per state energy should change from a particle to a wave. This requires a dynamic property of energy. Energy in the particle state should have the drive to probe energy in the distributed waveform, which is a state with less available energy.

Consideration 3:
The third consideration deals with the necessity of the duality of particle and wave having to be maintained, as long as a quantum system is active. A mechanism which enables the back conversion of a wave into a particle must therefore exist. Energy of information, activated by a fundamental demon (a model of thought), has to be involved in the back conversion of diluted energy of a wave into that of a particle with concentrated energy. This information is needed to account for and compensate entropy, which is present in the diluted energy of the wave. Only a dynamic energy can handle such dynamic energy exchange phenomena. A conventional "sleeping" energy would have no interest in work and change.

Consideration 4:
A fourth consideration deals with the difficulty of explaining time orientation and self- organization of matter starting from our present time-invertible physics. But time orientation and self organization are visible in our environment and there has to be a mechanism which allows and drives them. To overcome this problem, a fundamental time orientation is needed and the only reasonable way to accomplish that is via a vectorial, time-generating energy. Only where energy is converted do dynamic changes and time dependent processes occur. To better understand the problem we have to learn a little about how time orientation is presently handled in science as a purely statistical phenomenon.

8 Observations on entropic time orientation and information

In our present physics all fundamental physical laws were formulated to be time-invertible. For every solution of a time invertible natural law x(t), a time inverted solution x(-t) can be found. How then can a process develop only into one time direction as we experience it? It is well known that in classical physics a statistical time arrow is identified and defined in the direction of entropy formation, the formation of disorder. Time is considered to be a problem of many particles approaching this statistical disorder in the direction of an increase of disorder. This direction is accepted as the direction of time. Physics today uses time simply as a scale to order events. Time is considered to be an illusion, since the theory of relativity shows that different timeflows are possible in different moving reference systems.

a)

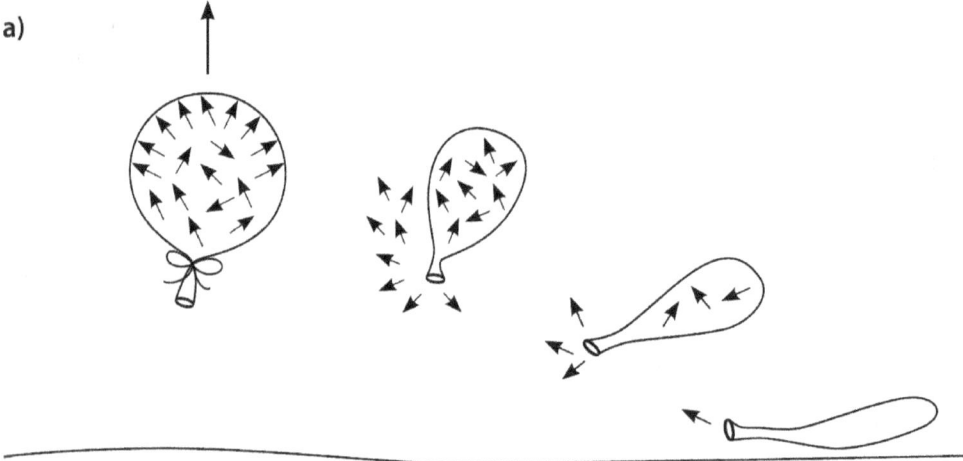

It is observed, that helium tends to escape from a balloon

b)

The inverse is never observed to proceed alone.

Fig. 13: Molecules of helium escaping from a balloon illustrate the phenomenon of irreversibility. In 1876 Josef Loschmid argued that due to time reversible physics, time inversion could force back the molecules. Today physics invokes quantum uncertainty and chaotic interaction in support of microscopic irreversibility and a statistical timeflow in one direction. Here it is proposed that time orientation is fundamental.

The increase of entropy (S), disorder, has received a statistical interpretation via Boltzmann`s famous relation S = klnW which relates entropy S to the natural logarithm of the number of possible micro-states W (with the Boltzmann factor k as the proportionality factor). Can an entropy increase also be derived from and be based on a purely energetic interpretation?

The classical entropy interpretation is that the more ordered a system is and the more information we can have on it, the smaller its entropy, or disorder, and vice versa. This notion was underlined by studies performed by the Austrian physicist and philosopher Ludwig Boltzmann, who lived from 1844-1906. He derived the so-called H-theorem, which appears to show that a time-invertible statistical ensemble finally develops into a time-oriented one (a phenomenon which defines the statistical entropic time direction). This has been interpreted as proof that entropy increase is a statistical phenomenon, a process due to the chaotic statistical activities of a large number of particles, which were originally subject to time-invertible physics. This statistical interpretation of entropy increase appears to contradict my interpretation that dilution of energy in space and time increases entropy on a fundamental level. It is, therefore, necessary to have a closer look at the derivation of the classical formalism.

This formalism considers a gas in which particles are subject to entirely reversible statistical collisions. Boltzmann now describes the behaviour of the time invertible particles via formulas and studies how their performance may develop. He does that by abandoning an initially very complex description of individual particles by simplifying it using a description of the particles in selected groups. The derivation of the H-theorem by Boltzmann is, strictly taken, based on a "one state" to "one state" replacement of deterministic dynamics by a "many state" to "many state" so-called Markovian mixing. It is a statistical procedure aimed at predicting future probabilities on the basis of little information on the past. It involves a random process, which can be characterized as memory-less. Memory and information is simply partially lost. The transition from the very detailed initial conditions to the latter procedure is, of course, mathematically associated with a loss of information. This occurs either by ignoring detailed correlations between particles, or between subsystems, or via the so-called coarse graining process. Here, a fine-grained earlier description is replaced by a coarse-grained one. What Boltzmann did not know, and what we already learned when explaining the Maxwell demon, is that information has an energy content (energy = kTln2 = 0.69 kT for 1 bit of logic operation, k = Boltzmann constant, T = absolute temperature) (Szilard (1929) (von Neumann (1966)). Boltzmann, in his calculation, has thrown away information, and thus energy.

The H-theorem and related more modern statistical mathematical calculations, which derive statistical entropic time orientation from time-invertible fundamental processes, thus throw away information (with an energy content) and thereby, of course, convert available energy into a non-available one. The system, after losing information in the form of energy, thus obtains an entropic time orientation, and can, for this reason no longer be inverted in time. It can no longer reach the initial condition. But this property is artificial, imposed via the statistical mathematical calculation procedure, which mathematically abandons information which contains energy. The system is obviously no longer the same after energy-related information has been abandoned and eliminated from the system. This remarkable fact, which is typically not mentioned in textbooks when the statistical, "entropic" time arrow is explained, underlines a present tendency of the neglect of the role information plays in fundamental physical science when described by mathematics.

Today everyone witnesses the dramatic impact of information processing and handling in our everyday life. It may play a similar role in nature and in the laws of physics. But it is not considered properly. For the present approach it will play a key role. Also the role of mathematical procedures for describing physics has to be examined more closely. Often mathematical derivations of physical situations are simplified to reach more elegant formulas. If information on the system is thereby thrown away, the simplified formula no longer exactly describes the original physical system. According to the quantity of information dropped, and since information has an energy content, the system described will now have an energy deficiency. It is no longer the same and may behave quite differently. Above all, the system is not reversible anymore. It cannot any more develop towards the past. It can develop only in the direction of the future, where it has been forced through mathematical manipulation. A Maxwell demon with information or energy from outside would be needed to invert the system again in the direction of a reduction of entropy. However, such a process would not change the time orientation. This is another piece of evidence showing that systems only develop into one direction, following a basic law. Meanwhile it has even been experimentally shown that thermal energy can be harvested from a thermal source via information (energy) input according to the concept of a Maxwell demon (Toyabe et. al.,2010) (As explained, this Maxwell demon re-establishes order from chaos through the input of information which has an energy content). No change of time orientation occurs. Time proceeds also in direction of entropy decrease.

When time orientation is manipulated by mathematical statistics to get the statistical entropy arrow what then generates time orientation which we actually recognize in our environment? My understanding is that energy conversion drives time. The changes which make up time are caused by energy conversion processes. Without energy conversion, nothing changes. But what time is it exactly and how can energy drive a fundamental time arrow?

9 The new energy theorem

The new energy property which I would like to introduce as a basic law is very simple:

"Energy has the tendency to decrease and minimize its presence per state, within the constraints of the system"   (7)

This minimization has to be understood to occur both in time and space, since increased space offers additional states that can be occupied. The new energy principle should now be explained using a simple example: a state mentioned here is the specific form which matter takes on for a temporal or permanent storage of energy. When a molecule such as the green plant dye, chlorophyll (fig. 14), absorbs energy in the form of a light particle, an electron is transferred (excited) from a lower into a higher energy "state". This state is a diffuse distribution of the electron within the chlorophyll molecule, which is different from the distribution of this electron before light was absorbed. In the higher energy "state" the electron may survive one fraction of a millionth of a second. Then it will again fall to a lower state, while releasing energy in the form of vibrations, rotations and (or) light with lower energy. This last process follows the proposed law that "energy has the tendency to decrease its presence per state". The energy, initially concentrated into a higher state, is redistributed into particles of light with lower energy and into vibration and rotation movements of molecules contributing to heat.

Since energy turnover generates changes and since changes take time, as we experience it, this tendency or property of energy generates time. What we see changing in time is what energy turnover generates. It is action, which is defined as energy multiplied by time. Action is a very important quantity in science, because it is addressed in a very fundamental principle, the principle of least action. Energy has, as fig. 14 depicts, two possibilities to decrease its presence per state. It can do it depending on time, and it can also do it by diluting its presence in space. Energy can, for example, leave the place in the form of heat, or it can leave it in the form of luminescent light. Energy will thereby be converted, which means that free available energy will become less available or non-available, for example in form of molecular vibrations and rotations, which define the moderate temperature of a thermal environment. This non available energy (entropic energy described by the product of entropy and temperature) may be present in the form of a reservoir with too low temperatures for energy recovery. This tendency of an entropy increase in a confined space is exactly what the empirical 2nd law of thermodynamics describes. In a closed space entropy approaches a maximum (see later). From both such considerations and earlier ones and since entropy means non-available energy, it may be concluded that it

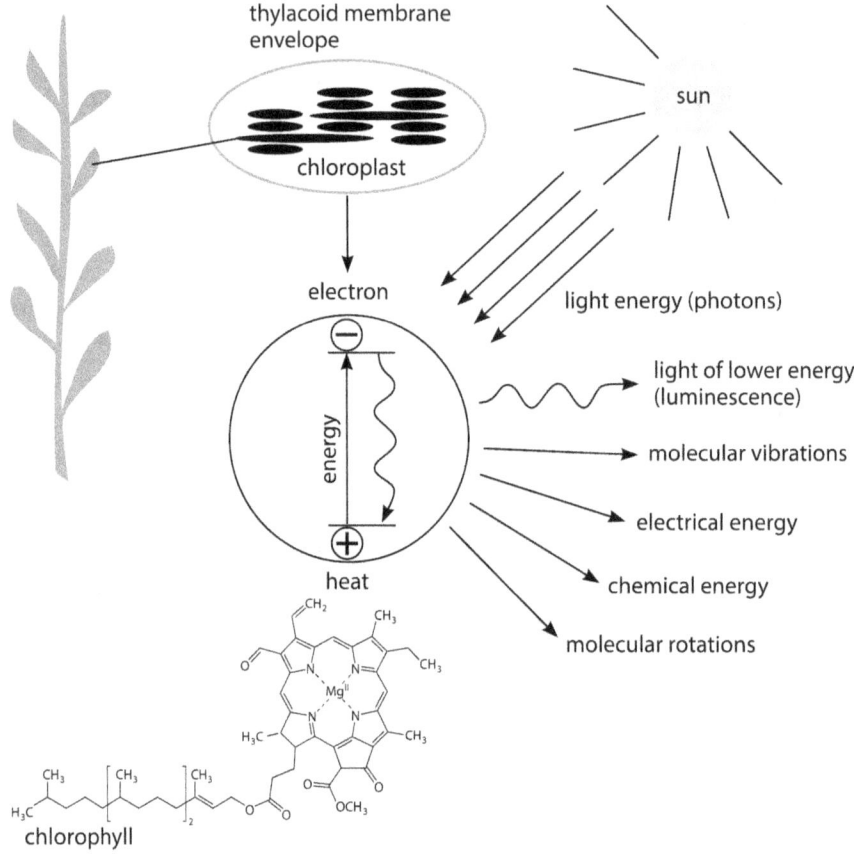

Fig. 14. In natural processes free, available energy is observed to decrease its presence per state in time and to spread into a larger volume. The example selected here shows a chlorophyll molecule in a green leaf, which after absorption of solar light transfers an electron into an energized state. From there its energy is redistributed generating a temporary electron on a lower energy state, but also a lower energy photon as well as kinetic energy in form of rotating and oscillating molecules in the environment.

should be possible to generalize the statements into a fundamental law saying that

"a dilution of energy density decreases its working ability"   (8)

That means its "entropic" energy content, the content in non-available energy, increases when energy is diluted in space. We will later see that this formulated fundamental property is exactly what quantum physics violates with the result being quantum paradoxes. How can we intuitively understand that diluted, distributed energy has an inferior energy value? Let us look at a charcoal burner. He needs a certain quantity of energy in the form of wood to build a wood pile, covered with soil, to heat the wood in the absence of oxygen for charcoal production. Two situations are depicted in fig. 15. In one case (a) the wood is already at the site of charcoal burning. In the second case (b) it is still distributed in the form of small heaps in the surrounding forest area. In the two cases the total amount of energy in the form of wood is the same, but the distributed wood has a lower energy value. Additional energy in the form of information about the localization of the small heaps and for transport will be needed to get this wood to the place where charcoal is burned. Many additional examples could be presented, which demonstrate that the same amount of concentrated energy is more useful than its equivalent in diluted form. If one, for example, puts a coffeepot into the sun, it will not boil, but by using a parabolic mirror one can get it to boil. Here a technical device, which incorporates information, redirects and concentrates sun rays. Information helps to reduce entropy.

a)

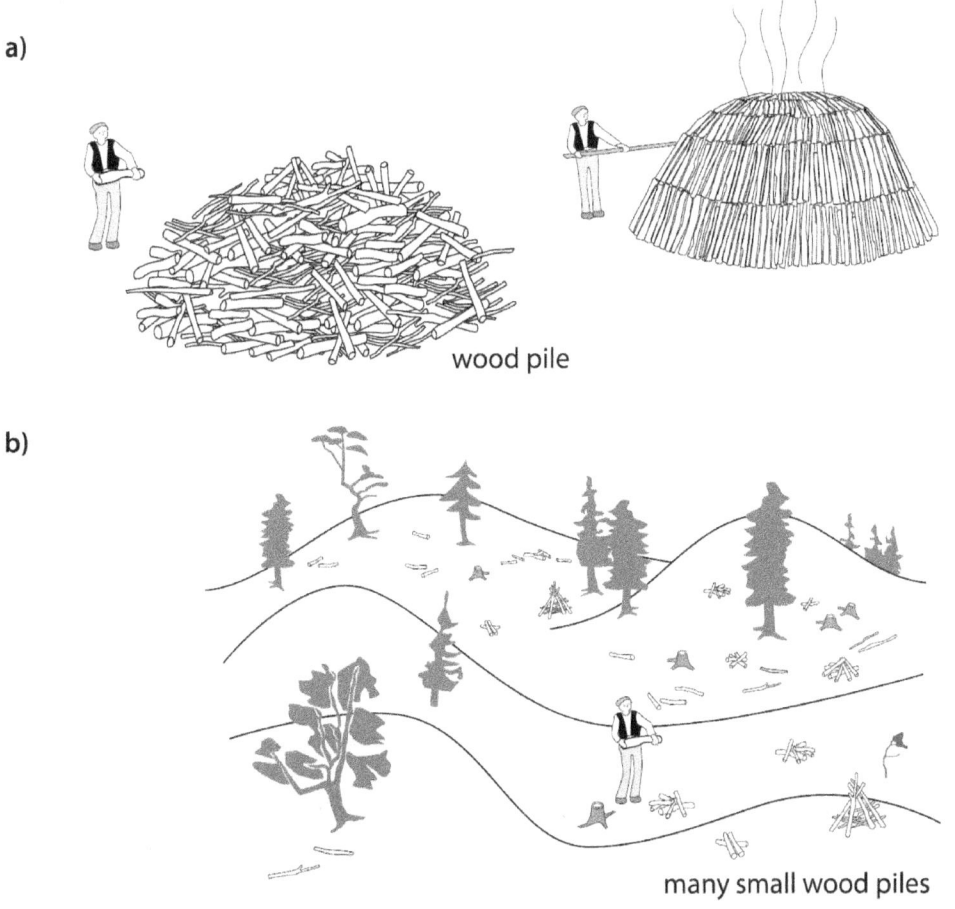

wood pile

b)

many small wood piles

Fig. 15a and b show a charcoal burner preparing a wood pile for charcoal production. In one case (a) the wood is already piled up near the future charcoal burning site. In the second case (b) the wood is still distributed in the surrounding cleared forest in small heaps. Why is the effective energy value of the wood in case (b) smaller? Additional information and energy is needed to collect it.

A process of redistribution of energy from a concentrated form to a distributed form of course only works if it is not forbidden by other restrictions. Such an energy system could be the quantum state. It can be represented by an electron or a photon. A photon can be described as a particle or as a wave. When absorbed by a small molecule it acts as a tiny particle. When diffracted by a lens, it acts as a spread wave. Let us assume the state of a particle is activated. Without other interaction possibilities the photon cannot, according to classical quantum theory, convert energy and react in time. It cannot produce action, energy multiplied by time. The photon, however, could attempt to decrease its presence per state. It could dilute its concentrated energy into the energy of a wave. In fact it has somehow to change between the particle and wave state because both properties are alternatively available. Depending on the type of radiation, the distribution of the wave can amount from fractions of a micro-meter, a millionth of a meter, to meters and kilometres. There is obviously a dramatic energy density change between a particle and a wave. Since physics discovered that a particle or wave representation of electromagnetic radiation are equivalent, and since energy tends to decrease its energy density (postulate (7)), it may be concluded that in a quantum state

"there is an ongoing interchange between the state of a particle and that of a wave".
(9)

In support of this claim it may be argued that a quantum, when arriving, does not know in advance whether it will encounter a small molecule for absorption or an extended layered structure for interference. A quantum must therefore be able to change from the particle form to the wave form and vice versa.

To summarize the proposal again: energy has the tendency to decrease its presence per state in time and space. In a quantum state, when energy turnover is not permitted, energy, in an attempt to decrease its presence per state, oscillates between the particle and the wave state. Energy probes the effect of space. Since the effect is a dead end road for a quantum state, the system again converts back into a particle. How this works and can be understood in more detail will be outlined further below.

But firstly it should be emphasized that the three statements concerning dynamic energy, (7), (8) and (9) have essentially become part of our new "symbolic form". The first one is the most important and claims an energy property aimed at decreasing its presence per state. The two additional ones are essentially deduced from the first and may help to understand nature in certain situations. They concern the loss of energy quality due to dilution of energy in space and the behaviour of dynamic energy in dealing with the particle-wave duality. It will be seen later that not considering such property in classical quantum theory is essentially responsible for irrationalities and paradoxes.

What I have defined as a dynamic energy theorem sounds reasonable, because dynamic changes and a distribution of energy into smaller quantities is observed everywhere in our environment as nothing changes without energy turnover. A fundamental timeflow related to energy turnover is therefore instinctively apparent. However, my statement and the new "symbolic form" nevertheless mean a paradigm change. Thomas Kuhn, an eminent historian of science, characterizes a paradigm as a set of practices that define a scientific discipline basically via universally recognized scientific achievements (Kuhn, 1970). There is presently no existing fundamental law in physics that defines a timeflow. The calculated statistical, entropic timeflow that aims at explaining our lived reality of changes is, as already explained, based on a mathematical manipulation, an abandonment of information on the described and calculated statistical ensembles. I, however, insist that a fundamental time arrow, as defined by (7), (8) and (9), is necessary to overcome irrationalities in fundamental science. I am even convinced that it can be deduced from classical physical experience in a straightforward way:

10 *Decrease of energy per state and principle of least action*

We should shortly stop here and reflect on the fundamental meaning of our dynamic statement on energy. Is it something exotically new or is there a convincing link to classical physics? Let us look at a stone rolling down a hill (fig. 16). It will follow a special path along which it will assume different values of potential and kinetic energy, that is of energy of position and movement. It is well known in physics that the path which the stone takes, follows the "principle of least action". It means that the time integral over energy E (which in this case is kinetic energy T, the energy contained in speed) minus potential energy V (the energy considering the position in the gravity field above ground E = T-V) approaches a minimum. The outcome is a number. In other words, the most efficient and energy-saving way for the transfer of the stone downhill is selected by nature. The principle of least action is very fundamental and many laws of physics can be derived from it. Examples are the classical equations of motion and the principles of Maupertius and Hamilton. The path of a stone thrown through the air, as well as the path of a planet circulating the sun are derivable from the principle of least action. Fermat`s principle for optics saying that light takes the shortest path is also based on it and it can also serve for the derivation of the general theory of relativity. The principle of least action is a fundamental mechanism which nature applies. As yet nobody can, however, explain why nature has adopted that principle and what it means fundamentally. Science is satisfied with the concept of a "sleeping" energy subject to the principle of least action. I, of course, am not and will show that this important principle explains much more than is classically expected:

Let us look again at the rolling stone. If the time frame of the "least action" integral for the rolling stone ($\int\limits_{t1}^{t2}(T-V)dt$ is subdivided into many small time integrals of energy, these also have to become minimal. It means that each infinitesimal small time integral $\int\limits_{t}^{t+\Delta t}E(t)dt$ over energy at time t has also to approach a minimum. This is mathematical knowledge listed in textbooks (e.g. Feynman lectures on physics, 1964). Let us now investigate what it actually means when these very narrow time integrals over energy should become minimal. The classical understanding of energy in physics is that it behaves as a "skalar" quantity. Energy "sleeps". A "skalar" is not an oriented, vectorial quantity, but simply a given number. I say that a given number cannot approach a minimum. It is given. In order to be able to approach a minimum, as the principle of least action requires, it has to have a property which is able to implement that behaviour. Energy has to be an oriented quantity, a quantity oriented in time, which is able to become a minimum. The time integral over energy in the principle of least action should obviously be understood as a non-equilibrium statement: the locally present energy E(t) aims at becoming minimized. Otherwise, as a pure number, it can not become a minimum.

And considering the three-dimensional geometry shown in fig. 16 of a hill from which the stone rolls, a minimization of energy should be approached in time and space. This is actually exactly the new dynamic energy postulate (7) proposed in the present study.

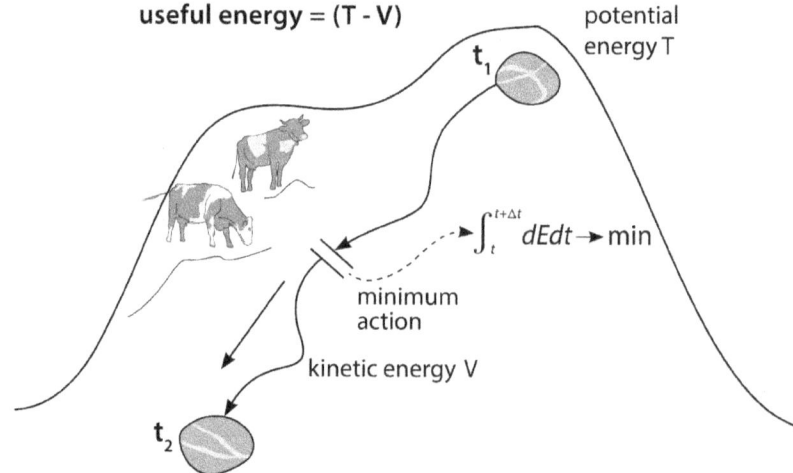

Fig. 16. A stone rolling down from a hill follows the "principle of least action". Nature selects the most efficient path in time and space. Important laws of physics are derivable from this principle. Since any tiny portion of the "least action" integral has equally to become minimal, a statement on the time orientation of energy itself follows as a necessary consequence. It is the "new" energy paradigm (7) proposed here.

The paradigm change for energy introduced here is therefore nothing more than a dynamic interpretation of the principle of least action applied to an infinitesimally narrow time period.

This makes it a statement which is compatible with the most fundamental basis of the understanding of physical mechanisms. Such a conclusion can also be confirmed with another argument: action, energy multiplied by time, is the consequence of causes. And such a consequence needs a timeflow. Action is consequently linked to timeflow, however not in classical physics. Neither energy nor time are oriented and linked to a timeflow in classical physics.

Just before, it was mentioned that nobody knows why nature has selected the principle of least action as a basic principle. It was also shown that the dynamic energy postulate proposed here can be derived from this principle, and that energy has to be "dynamic" and able to minimize itself. Now it has been pointed out that the principle of least action can even be recognized to be equivalent to the dynamic energy postulate. It can be derived from it. In order to find the minimum of the least action integral, which the principle expresses, one has to differentiate it. According to the Leibniz rule, which has to be applied for this aim, the "dynamic" energy remains and has to minimize its (free) energy content. This confirms that my "dynamic energy" statement is entirely reasonable. And it conduces to another astonishing result: the principle of least action, the principle which has become so important for classical physics, expresses nothing other than the existence of a dynamic energy and of an oriented time in nature. For the first time in the history of science a reasonable explanation for the fundamental significance of the principle of least action is given. And it is an entirely logic explanation. It is simply the statement, that nature is fundamentally dynamic, and that an energy-containing system minimizes its free energy content and drives time, the trace of changes. This is more or less what the ancient philosopher Heraklitus already observed (fig. 11) and what we are also seeing all around us. The principle of least action, in physical terms, declares exactly this fact. Nobody before seems to have seen such a simple truth behind the mysterious principle of least action.

The reason why the fundamentally dynamic nature of the principle of least action has not been recognized and valued before in such a context in traditional physics may have two important reasons. First, energy is classically defined as a scalar quantity, a number, with no interest in inducing changes. And secondly, all fundamental laws, even those derived from the dynamic principle of least action, were understood as entirely time-invertible. The time-oriented mechanisms hidden in the principle of least action were simply not recognized and exploited, because they appeared to contradict reversible and time invertible properties of classical quantities in physics. A dynamic energy principle, which the principle of least action is, was applied to time-invertible physics and energy values in form of numbers. On the basis of these considerations, one also understands the manipulation which mathematical physics has been carrying out to derive important theories. Einstein, for example, used much effort to formally derive the general theory of relativity from the principle of least action, but he used the classical notion that energy is a scalar, a number, and time is a simple parameter monitoring change and he only considered time invertible physical laws in the theory. Orientation is thus excluded in derived mechanisms. Nevertheless, the general relativity theory is now used to describe tremendously dynamic hypothetical processes such as the Big Bang or inflation, an explosive expansion of empty space, just by changing parameters or numbers in the theory, and looking at the outcome in figures. It is time to critically re-evaluate such unrealistic assumptions. This will be attempted later.

Via a dynamic interpretation of the principle of least action, also the understanding of a fundamental time flux proposed here becomes evident. The reality of change is action in its dynamic, time-oriented form. The integral over many small elements of action provides a sequence of action from which time can be separated, when action is divided by the energy turned over. For this an appropriate mechanism is needed. It occurs in a clock or in our brain. This way a calibrated measure for changes is obtained, which is time. However, it is artificial, a tool for measurement and calculation, a tool for measuring changes. Time has no substance, it contains neither matter nor energy. It can therefore not directly be measured like flowing water, or the blowing wind. It has to be filtered out from action with the help of a special machine, a clock. Action comes from the movement of a pendulum, the oscillation of a quartz crystal, or the energy transition of an atom. What we get is simply clock-time, as Einstein uses it in the theory of relativity. Classical time is not oriented, but simply a calibrated ordering parameter. This is obviously the reason why contradictions are experienced in some theoretical models of the general theory of relativity, which consider the principle of least action. Time is classically used independently of dynamic action, which has its limitations, as we will see later.

After deriving it from a very fundamental physical principle we will, in the following, attempt to use this dynamic energy property or the newly created symbolic form, to get rid of irrationalities in science theories. It should create our new reality of understanding. By applying it, nature should become better understandable to us.

In the last two paragraphs we searched for relevant laws (stated as (7), (8) and (9)), which could serve as plausible new "symbolic forms" to better understand fundamental physics. This followed advice given by science philosopher Karl Popper (1978): "If you want to recognize, then you have to search for laws". These laws should now be applied.

11 Thermodynamic laws are consequences

Since time orientation and a timeflow is actually observed in nature, entropy formation should thus have a fundamental energetic reason and is rate controlling for dynamic phenomena. The paradigm change for physics proposed via relation (7) claiming that the presence of (useful) energy per state of a system is decreasing and minimising in time and space, accounts for that. In the last chapter it was shown that this statement is nothing new, but equivalent to the recognized principle of least action if it is interpreted in a consistent way, which is dynamic. This can and should actually be adopted as the fundamental basis of time- orientation and of the empirical 2nd law of thermodynamics, which states that the entropy, the non- useful, "chaotic" energy, in a closed system approaches a maximum. Up to now the second law of thermodynamics has had to be considered as a purely empirical law. It followed on from many experimental observations, last not least from drilling canons, which released heat into the surroundings. It, however, turned out to be impossible to derive this 2nd law from Newton`s laws of mechanics. This is in fact a catastrophe about which physicists do not like to talk, and it is related to the fact that Newton`s fundamental laws are entirely time invertible. They do not care about time orientation, but the second law of thermodynamics does. It is empirical and definitively correct. And this 2nd law follows directly from the "dynamic energy" idea proposed here (relation (7)). And since the dynamic energy property is claimed to be fundamental, and mathematically derivable from the principle of least action, it is also applicable to elementary and molecular processes. And this will turn out to be quite important for eliminating paradoxes in quantum physics.

It is consequently not seen as a contradiction to replace the mathematically "manipulated" and enforced statistical H-theorem of Boltzmann to explain entropy increase (as discussed in chapter 8) with the postulated fundamental new energy law and have exactly the same consequence: the energy`s tendency to decrease its presence per state in time and space (discussed more in detail in (Tributsch, 2008)). The result is entirely identical and explains the decrease of the working ability of energy when responding to time or space, but on a fundamental basis. This not only allows but requires the application of this energy concept also to quantum processes. Most importantly, it simultaneously identifies a fundamental, energy-related orientation of time, in contrast to the presently established statistical entropic one. We will have to discuss the new suggested meaning of time in more detail further below. As energy conversion and time are linked in the form of action, energy multiplied by time, time is no longer an independent ordering parameter as we use it in classical physics. Timeflow has to be calculated from sequences of action, which, in my opinion, are the only time-related fundamental reality in nature. By dividing the flow of action by a well-defined flow of energy we arrive at a timeflow. However, time itself, having no substance, cannot be directly measured. When this timeflow is properly calibrated with respect to the daily and yearly cycles in our environment, we obtain the clock-time. In classical physics the arrow of time is associated with the direction of entropy increase, the increase of disorder, or chaotic energy. There it is understood as a time-oriented statistical process, but I mentioned that its derivation, by Boltzmann, was based on an abandonment of information, which itself already imposes a time direction (chapter 8). It is only a superficial coincidence, based on the fact that a turnover of energy in the form of action generates entropy, but entropy generation may vary both locally and temporally. In reality there may be an entropy decrease due to the generation of order in one location, and an accelerated entropy increase in another close by. Time advances in a more definite and accurate way than the increase of entropy, because the continual flow of action created by clocks is recalculated by dividing it by a constant energy flow and it is calibrated with stable astronomical time cycles. It is not the statistical entropy which increases, but a controlled generation of action and its calibration is the basis of time measurement.

The following statement can be made:

"Time is the track, which energy turnover carves into the environment or our brain by generating sequences of action". (10)

This reinterpretation of energy properties and the adoption of the altered concept of time indeed automatically results in the 2nd law of thermodynamics (Tributsch, 2006), but, as already mentioned, on a fundamental basis. This recognition of a fundamental time and space-oriented energy property (Tributsch, 2008), however, enables and requires the extension of this fundamental energy behaviour to quantum phenomena. This way an application of the created new "symbolic form" to elementary particles and quantum processes becomes possible and necessary.

My definition of a dynamic energy not only explains the empirical second law of reversible thermodynamics, but also becomes a straightforward basis of irreversible thermodynamics, the field of science dealing with irreversible processes, which also include life itself. The transition from reversible thermodynamics, where energy quantities are so called "quantities of state" to irreversible thermodynamics, where they become time oriented and dynamic variables, has been a challenge for many decades in science. Great efforts were undertaken to explain microscopic instabilities of time-invertible processes as sources of macroscopic instabilities that generate statistical timeflow (Nicolis and Prigogine, 1977)(Antoniou and Prigogine, 1993)(Prigogine, 1980, 1984). In order to "derive" the law of minimum entropy production, which governs the linear, simplified range of irreversible thermodynamics, complicated mathematical manipulations already had to be performed. For example, individual trajectories of particles were grouped into ensembles of trajectories. Under appropriate conditions these ensembles of trajectories developed into chaos, which no longer allows the system to return to the initial conditions. It is the so-called "symmetry breaking" reaction: small fluctuations act on the system, which crosses a critical point, which decides the system's future fate. A new branch of a "bifurcation" thus follows, from where the system does not return. Irreversibility is obtained.

With these discussed attempts to derive time-oriented behaviour from time invertible physics, again information, and thus energy, was abandoned by advancing from individual trajectories to ensembles of trajectories. With information, however, energy is thrown away from the system studied. This changes the system via the applied mathematical manipulation. It is not the same system as before the manipulation. There is, in addition, another serious problem with such a strategy for generating time orientation. For the transfer of a system into deterministic chaos, which opens a route without return, time orientation would already be required before the chaos appears. There is simply no deterministic chaos without feedback processes. A simple example of feedback is given by the self-regulation of a heating system. Feedback requires a distinction between "before" and "after", and such a distinction is possible only in the presence of a timeflow, a time arrow. When the room temperature falls below a certain temperature, then the thermostat reacts by restarting the heater. In a theoretical calculation quantities obtained from a given algorithm have also to be allowed to again interfere with the original, starting quantities to yield a sufficiently high degree of complexity for describing chaos. The relevant equations have to become non-linear, which means that the outcome, when plotted on the basis of initial quantities in a graph, must deviate from a straight line. Starting from a time invertible equilibrium world this is not possible without mathematical manipulation.

All these sophisticated mathematical problems disappear with the dynamic energy hypothesis proposed here and in part discussed previously (Tributsch, 2008). Energy is already a dynamic variable. And dynamic energy drives a time arrow with entropy production becoming also a dynamic variable. Feedback processes are possible and, as a consequence, a self-organisation of matter can occur. The door towards irreversible thermodynamics, the science, which is programmed to describe life and all the changes around us, is entirely open. It should be pointed out here that up to this point the time concept has been applied in a rather relaxed way. As previously mentioned, energy, which drives time, does not generate time in its mathematically pure form, but in the form of elements of action, which are described as a product of energy and time. Ongoing action is what we recognize as changes in our environment. A pure quantity such as time should, as already mentioned, not exist in nature. Only action exists. However, a timeflow can be filtered out from a flow of action. When action is divided by the energy turned over, a quantity, which we understand as time, is extractable. As a consequence, time periods are experienced in our brain and by time experiencing organisms.

Let us now return to the meaning of irreversible thermodynamics, which describes processes proceeding far from equilibrium, such as life, the vortexes in a waterway, or a hurricane.

We recall that reversible thermodynamics, with its 2nd law, states that a maximum entropy is reached in a closed system. Where does irreversible thermodynamics aim in physical-chemical processes, which are open systems, so that energy can flow through? Is a statement on entropy possible, which explains where a system, exposed to (non linear) irreversible thermodynamic conditions, will eventually go? The "symbolic form" of the dynamic energy hypothesis presented here allows a straightforward conclusion: if sufficient energy is involved, the energy systems, e.g. a weather condition, or an ecological environment, will develop towards a situation of maximum entropy production. All the energy introduced into a system which decreases its presence per state will finally end up as "non-available" energy. At what rate will this happen? This should depend on the constraints of the systems. Some systems will have slower pathways to decrease the presence of energy per state, some faster ones. When one removes constraints, entropy production will increase. The nonlinear range of irreversible thermodynamics allows self-organisation of matter. Processes of this kind are faster in turning over energy than ordinary, much simpler processes since they involve autocatalysis and feedback mechanisms. If they develop, they will compete amongst each other for the available energy so that the rate of energy consumption and entropy production will gradually maximise:

"Far from equilibrium, entropy production (and parallel build-up of order and information) will approach a maximum rate within the constraints of the system"

(10a)

Our "symbolic form" analysis has thus also yielded a thermodynamic criterion for the non-linear range of irreversible thermodynamics. We now have an idea at which destination a phenomenon such as life is aiming.

Does such a conclusion make sense? On the basis of empirical observations a similar result has already been reached by other scientists, but they could not derive it from more fundamental laws and prove it. The first considerations on maximum entropy production seem to go back to Jaynes (1957) who studied it in context with statistical processes and defined it as the most uninformative distribution of information. A maximum entropy production rate in the atmosphere, as a state determining the climate, has been proposed by Paltridge (1979) and later others (Liu et al. 2011). Pioneering work on maximization of entropy production rate in relation to human ecology came also from Swenson starting from 1988 (Swenson, 1997). Maximum entropy production within the constraints of a system far from equilibrium can be understood on the basis of a simple example. A person, a self-organized system, out in a winter storm loses heat through his clothing. If the coat is taken off, a constraint is changed and heat loss is increased. More entropy production occurs. If the person undresses further, a further increase in entropy production follows. The system always tends to maximise it.

The living activities of man are critically determined by energy conversion. Since the time of primitive existence from nomadic life man has continuously changed the constraints for energy turnover. He now turns over 60-120 times more energy than a primitive man. And man is inclined to turn over even more energy if it

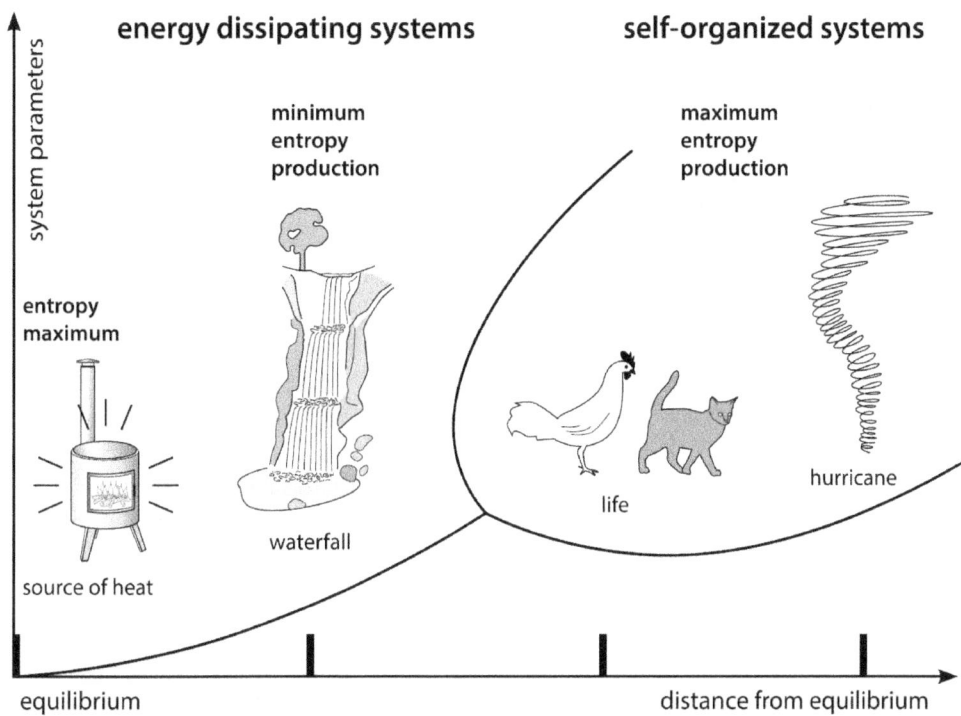

energy dissipating systems   self-organized systems

system parameters

minimum entropy production

maximum entropy production

entropy maximum

source of heat

waterfall

life

hurricane

equilibrium   distance from equilibrium

Fig. 17. Physical-chemical systems follow different entropy laws depending on their distance from equilibrium. A closed system in equilibrium develops towards maximum entropy. Close to equilibrium a minimum entropy production is found. Far from equilibrium, where life functions, a maximum entropy production is approached, according to considerations presented here.

becomes available to him, that is, if the constraints allow it. This principle of maximum entropy production, derivable from empirical observations, and following here from the "dynamic" energy postulate, has, however, found strong opposition. It was argued by John Ross from Stanford University that the rates of chemical reactions are controlled by (non-vectorial, non-oriented) thermodynamic quantities such as Gibbs free energy and activation energy and not by the rate of entropy production (Ross, 2008). And in consequence, he speaks of the "invalidity of the principle of maximum entropy production". The criticism expressed by John Ross is perfectly reasonable, but it is reasonable only on the basis of reversible thermodynamics, its non-dynamic quantities of state and its traditional understanding of kinetic mechanisms. This situation is entirely changed with my proposal to introduce a dynamic property of energy (derivable and justifiable from the principle of least action). When energy is given the property or tendency to decrease and minimize its presence per state, then it simultaneously drives the rate of entropy production and this leads straight to maximum entropy production far from equilibrium within the constraints of a system. The criticism forwarded by J. Ross no longer applies and, in fact, the requirement mentioned by him exactly matches the introduced paradigm change. The free, available energy no longer "sleeps" but drives entropy production. The law of "maximum entropy production within the constraints of a system" for the non-linear range of irreversible thermodynamics is a straightforward consequence of the "dynamic" energy proposal. It is supported by a great deal of empirical evidence. It also explains the destructive potential of atmospheric storms, and why they can become even more dramatic on other planets. This success is another indication that the dynamic energy approach is on the right track.

Later, in chapter 27, I will be talking about biological evolution, where self-organized systems compete for survival. There, on the basis of additional arguments, the conclusion is reached that biological evolution is aiming at maximising energy turnover. Maximum energy turnover, of course, generates maximum entropy turnover. This is exactly the conclusion reached here for open systems functioning far from equilibrium. Living systems following evolution are simply subject to the basic law identified here. They approach it provided the restraints of the systems allow it.

12 How did quantum physics become irrational?

After getting such a positive impression of the applicability of the proposed new symbolic form to irreversible thermodynamics and self-organized systems such as life, let us now continue looking at the challenge of irrationality in quantum physics. We have reasoned that quantum physics may have a problem with missing information for the observer. Quantum physicists deny this on the basis of Bell`s theorem (Bell, 1964). This theorem basically states that no physical theory of hidden, newly added variables can reproduce all predictions of quantum mechanics. In other words, it essentially insists that irrationalities in quantum physics are fundamental. However I argue that quantum physics has a problem with the "symbolic form" used to understand it. The symbolic form, applied up to now, is programmed not to know that dilution of energy changes its properties. The way in which one classically looks at quantum systems via the "symbolic form" given by the Copenhagen school does not allow rational cognition. Anyway, we do not want to copy all predictions of quantum physics, including its paradoxes, but we want a cleared-up, understandable version, which is still fully compatible with experiments.

From where, by the way, does the strange property in conventional quantum physics of an energy which is unaffected by dilution in space, come? This is a very important question, but fortunately relatively simple to answer. In an effort to explain quantum phenomena both in a particle and a wave picture, quantum physicists claimed that a presentation of a system as a particle is equivalent to its presentation as a wave. Some quantum phenomena can be understood only in the particle picture, such as photon absorption by atoms or small molecules. Here atoms or molecules absorb the light of a wavelength which is easily a thousand times larger than their own size. How could light energy be focussed dramatically fast during absorption if not by an already existing light particle? Other optical phenomena, such as interference colours, for example of oil films floating on water, can be explained only in the wave picture. Depending on the thickness of the layer, which is generating this interference, different wavelengths of light are selectively superposed or eliminated. Physicists insist that the particle and wave aspect of light are equivalent.

What this duality of particle and wave actually means is visualised in fig. 15a. A particle concentrates its energy $E = mc^2$ into one point of the size of the particle. Within a wave, on the other hand, its energy is distributed over an area of the dimension of the wavelength. This wavelength can, depending on its energy or, which is equivalent, its frequency, reach macroscopic dimensions of millimetres, meters or kilometres. By identifying a point like particle with an extended wave, in other words by claiming particle-wave identity, quantum physicists have simply eliminated space with its influence on energy properties. Energy, expanding from one point (a particle) into space to represent a wave, simply remains the same in quantum theory. It thereby does not lose part of its capacity to do work, even though a point-like particle may extend to a wave of a length of micrometres, meters or kilometres.

The energy laws in classical mechanics and quantum mechanics are in this way to a significant degree contradictory. In contrast to classical physics and reversible thermodynamics space in quantum mechanics is not relevant for energy laws. Science uses contradictory "symbolic forms" for shaping intellectual perception in physics of neighbouring fields such as quantum physics and classical physics, as expressed in relation (3). This must necessarily lead to a certain degree of irrationality encountered during the attempt to understand the proceeding mechanisms. It arises as a consequence of contradictory symbolic forms through which we attempt to gather cognition. It is simply proposed here that the identified dynamic energy behaviour of losing working ability in space and time (Tributsch, 2008) as a fundamental law of nature should also be used for quantum physics. In a future elaborated theory such an energy property has to be considered as a vectorial (oriented), dynamic property of energy which responds to space and to time. When diluting into space, energy changes by losing part of its ability to work. And when energy is being converted and redistributed, it also changes by losing some of its ability to work, thereby creating the flux of time (Tributsch, 2008).

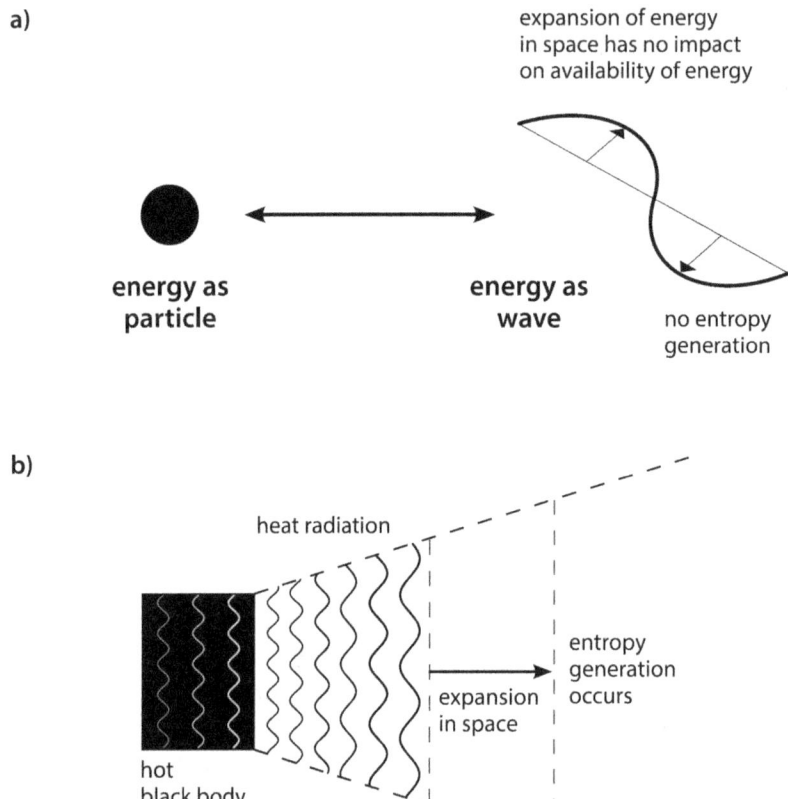

a)

expansion of energy
in space has no impact
on availability of energy

**energy as
particle**

**energy as
wave**

no entropy
generation

b)

heat radiation

hot
black body

electromagnetic
energy

expansion
in space

entropy
generation
occurs

Fig. 18. By stating a dualism of particle and wave (energy concentrated and distributed respectively) classical quantum physics has eliminated the relevance of space for energy (a). From classical physics it is however known that when radiation expands from a hot black body, non-usable energy in the form of entropy is generated (b).

As I already mentioned, the principle of causality, the observation that there is always a cause for an effect or a consequence, has shaped understanding of science since the time of ancient Greek philosophers. It is not difficult to see energy laws dominating the principle of causality. Without energy conversion there is no change. Typically, the change is observed distributed in space and time. Even when a distant event is generated through the transfer of information, for example, via a long distance phone call, energy conversion is involved. Information involves energy as well as any generation, transmission and turnover of information in space and time. It is consequently the sequence of energy conversion processes along a chain of effects, which shapes causality.

The empirical mathematical formalism, which quantum physics developed, did not only ignore energy properties in relation to space as discussed above (fig. 18). It also eliminated time, but time is required for causality to distinguish between before and after. Quantum phenomena, however, do not distinguish between before, present and after. Quantum correlation phenomena themselves are timeless, since they represent states which are not involved in energy conversion. For phenomena without relation to time, causality is, of course, not defined and has to be interpreted in an extended context of cognition. However, since energy is finally exchanged between quantum states, the introduction of time into the quantum formalism was necessary. The time-dependent Schrödinger equation of quantum theory obtained time as a parameter via the classical perturbation theory. It is the time of classical physics, clock time, which has been introduced here. Quantum physics tries to combine timeless and time-related processes.

With such peculiarities of energy properties in relation to space and time, one should not be surprised that the classical causality concept is no longer generally applicable to quantum phenomena. But later it will be seen that with the fundamental interpretation of energy dissipation and of time, a deterministic interpretation of quantum physics is possible.

In this context it is worthwhile having a second look at Bell's theorem which claims to prove the completeness of quantum physics (Bell, 1964). Quantum physics can apparently not be extended nor better explained via the introduction of additional, so-called "hidden" variables. Our understanding, however, is: a theory which is based on partially incorrect symbolic forms E_2 (ignoring the relevance of space for energy) and described by $P_2 = f(E_2)$, cannot be proven from within to be correct. It can only be demonstrated that it is consistent, and not improvable through modifications within the assumed, only partially correct, frame of symbolic forms E_2. But the cognition reached can still be fragmentary and incomplete, as seen from relation (3). This means that we should not allow ourselves to be intimidated by Bell's theorem, even though physicists claim that quantum irrationalities are proven to be fundamental. They are only fundamental within the "symbolic form" applied to look at them. And this symbolic form erroneously insisted that space is not relevant for energy.

13 How can quantum irrationality be overcome?

We have thus arrived at the conclusion that classical physics and quantum physics are based on the assumption of partially different energy properties. Expansion into space affects a relevant energy property, the ability to do work, in the first, however not in the second case. Our understanding of physics is thus based on relation (3). We perceive neighbouring fields of science, classical mechanics and quantum mechanics via, in part, contradictory symbolic forms. This may explain why we are faced with irrationality and counter-intuition in one of the fields, in quantum physics. We are trying to understand physics with two different and partially contradictory symbolic forms (a specific property of fundamental energy laws is different). This analysis definitely suggests that the cognitive problems with quantum physics are man-made. It is not nature which is irrational, but our mechanism for understanding nature is the problem. Quantum theory provides an apparently functioning practical tool (according to Healey, 2012), because it has been adjusted to function that way, but it is not complete in terms of information, caused by the only partially correct energy concept applied. In order to correct this there is a straightforward solution. We have to arrive at identical fundamental energy concepts, that is we have to make $E_2 = E_1$ in relation (3). Since classical physical understanding does not lead to rational contradictions, we should adopt the established concepts for its energy properties for all physics. I proposed to do this formally via a redefinition of energy properties for the particle – wave dualism, since dynamic energy will attempt to expand into space and will oscillate between the particle and wave form. The change in volume experienced for energy by the quantum system is thereby significant.

If one would allow energy expansion and dilution into space via quantum mechanical processes to occur without some parallel loss of working ability, contradictions to these empirically verified facts from thermodynamics and experience with electromagnetic radiation (as outlined above) would occur. However the reversible nature of the energy conversion process within the quantum phenomenon (no energy is finally turned over) also has to be considered (see relation (11) below). Energy in quantum systems can expand and it also loses working ability while generating entropic energy The formalism for the quantum system, however, should also set aside energy in the form of information to guarantee reversibility in the absence of energy turnover. A kind of fundamental Maxwell demon is needed to bring back the energy distributed in a wave to the shape of the particle. Since this demon will need information, and thus energy, to do so this energy has to be set aside by the particle before converting into a wave. It has to be set aside to act from the outside of the expanding energy system.

The energy of the particle and the wave should therefore not be identical and not ignore space, as seen in fig. 18a, but the energy of particle E_p should be equivalent to the distributed energy in the wave $\sum_w E_w$ (describing the sum over all w tiny energy particles) plus the fraction of energy which is no longer available for work E_e due to the expansion in space, and contributing to entropy, plus the "negentropic" energy E_n (energy in the form of information), which would be required to make energy conversion reversible (see fig. 19b). This would work via a fundamental Maxwell demon. He uses information with energy content E_n to handle the energetic tasks for the recovery of the distributed energy in the wave to reshape the original particle with energy E_p (h = Planck constant of action, v = light frequency):

$$hv = E_p \longleftrightarrow \sum_w E_w + E_e + E_n \qquad (11)$$

In other words: the energy in the particle E_p converts into the distributed energy of the wave $\sum_w E_w$ plus the non-usable energy in the form of entropy E_e plus energy E_n set aside in the form of information needed for the reconversion into the particle. The latter energy E_n is a kind of "information self-image of matter".

No energy is exchanged with the outside and the energy of information which is set aside from the beginning, is tailored in such a way that the total energy as expressed in formula (11) is sustained. We have thus permitted that the energy converted from a particle into a wave loses some ability to perform work (it assumes a microscopic form of entropic energy), but simultaneously provides energy in the form of information (also called negentropic energy because it behaves and acts somehow in the opposite way to entropic, chaotic, energy), which subsequently reconverts the entropic energy. The information provided by a hypothetical microscopic Maxwell demon is used to re-concentrate the energy into a particle (fig. 19b). This does not contradict the second law of thermodynamics, which states that the entropy increases in a closed space or volume, because information is assumed to come from outside. The system works as it does with information provided to a three dimensional printer, which assembles a three dimensional object via pure information: this is required to provide reversibility of particle-wave inter-conversion within the particle-wave duality. When the particle contains sufficient energy to account for the information needed for reconstruction, then the back conversion from a state of distributed energy plus the accompanying entropic energy can indeed proceed according to relation (11).

It may be asked here how inter-conversion between particle and wave actually works. Later (chapter 20) it will be seen that energy and matter can self-organize, because time orientation is found to be fundamental. With self-organized systems one ordered structure can be converted into another one just by a small change of parameters.

The information self-image of matter postulated here is related to energy (E_n), which has, if it is a real phenomenon, somewhere to exist. It must be detectable in nature around matter. This will, of course, be a touchstone for the model. Further below it will be shown that gravitation has all the attributes and properties of this kind of fundamental dynamic information quantity. It is recognized here as a self image of matter in the form of information. This interpretation (fig. 19b), of course, is in strong contrast to the classical quantum mechanical understanding of the particle-wave dualism. (fig. 19a).

a)   traditional model

particle

wave

b)   new model

particle

information
self-image + wave + entropy

Fig: 19 a) traditional model: in quantum theory particle and wave are energetically equivalent. b) proposed new model: when a particle changes into a wave, the distributed energy has less working ability and in compensation entropic, "chaotic" energy is generated. In order to guarantee reversibility, part of the particle`s energy is set aside from the beginning to supply information to a fundamental "Maxwell demon", which reassembles the particle again from the wave.

A few words about information: we are now living in a period, in which we can witness a very dynamic expansion of information technology. It has already affected nearly every aspect of life and civilisation. We are experiencing what information is able to do, but what is it physically? Information is neither energy nor matter, but it needs both free and available energy to be created and communicated as well as matter to be materialized. Information is also additive, or, if contradicting, subtractive. The energy of information E_n is proportional to the information or "negentropy", negative entropy (Brillouin, 1951). Information can be created and destroyed. But in contrast to energy information is not conserved in an isolated, closed system. Consider a room of a villa in ancient Pompeii, full of script rolls, covered and isolated by volcanic ash. At the beginning it contained a great deal of information. Then, during many passing centuries, the script roles degraded and the information disappeared. Energy, however, was conserved, but turned into a chaotic, non-usable form while entropy became a maximum. Even though information is so powerful in shaping technology, the concept of information has barely entered basic science and fundamental physical laws. This is very strange and indicates that something essential is missing from our understanding of nature. Why should nature have avoided something in basic laws, which turns out to be so powerful and useful for technology in our civilization? In fact, we are simply applying natural laws to handle information technology. Why should nature itself not be doing it? With the above formula (11) such a fundamental consideration of information for the functioning of natural phenomena is implemented here. In quantum processes nature thus generates and employs information-images of herself, of matter. They accompany the proceeding quantum mechanisms. It will be seen that these information self-images, expressed in energy term E_n in (11), are the key to overcome irrationality and paradoxes:

"Reality, matter and energy, and information on this reality, must be considered jointly to obtain a rational understanding of quantum phenomena" (11a)

In the presence of matter, which is subject to the particle-wave dualism, this self-image in the form of information will always be around. The energy involved in this postulated information, E_n, negentropic energy, must therefore show up somewhere in nature, when larger masses are present. This is a necessary consequence of the fundamentally dynamic property of matter and energy proposed here (equation (11)). Later (in chapter 20) it will be shown that this information - "substance" is gravitation. Identifying it with the information self-image of matter will later help us to understand some fundamental aspects of gravitation and of the dynamics of our universe. Information will be recognized as a key factor for understanding both elementary particles and the universe.

It should be mentioned here that intuitively some scientists have already speculated that information may play a role in quantum physics. Bohm, with Hiley (1993), for example, tried to associate incomprehensible quantum properties of elementary particles with information which these particles can read from associated fields. The particles were expected to have a "rich and complex inner structure and a rudimentary mind-like quality". The associated field guides the particle, for example through a screen with one or two slits respectively. Information has been invoked also by Zeilinger (2007) in relation to quantum correlation.

It is recalled that in the picture proposed here energy is no longer a scalar, a number, but becomes a dynamic, vectorial quantity, a quantity which responds to space, and in the case of energy conversion produces action and drives time. During energy conversion, when the quantum state is abandoned or changed, the energy content of a system is reorganized to yield an effective entropy increase (e.g. manifested in terms of heat, vibrational energy distributed over different states and in space). A time-determining process has occurred.

This paradigm change for energy (from a scalar, inactive quantity to a dynamic, vectorial one as expressed in (11)) has far-reaching consequences. And it better matches the quantity (energy), searched for since antiquity, which is conserved within all the change observed in our environment, than our present concept of an energy which has no relation to change nor interest in doing work. The resulting concept has been elaborated (Tributsch, 2008). In this work I feel I have also given sufficient evidence that these corrections to energy and the quantum mechanical particle-wave duality do not contradict experimental reality. One can go on using the established empirical quantum mechanical formulas and the Schrödinger equations, but the new symbolic form (statements (6) to (10)) allows us a rational approach. In addition, it will also allow a more detailed interpretation and understanding of quantum mechanical mechanisms.

Especially in relation to the information mechanisms, which I claim are involved in quantum physics and which are expressed by the quantity E_n in relation (11), new insights can be expected. One of these is related to understanding quantization, which historically has been found and described empirically. Max Planck attempted in 1900 to derive, via a mathematical formula, the radiation emitted from hot, black objects and found that energy is emitted in form of quantized portions. It is now well known that atoms and molecules accommodate electrons around the nucleus or the nuclei (when talking about molecules) on well-defined, quantized states and orbits, where these electrons find stable positions. At the beginning of quantum theory this was in no way obvious since an orbiting, negatively charged electron was expected to lose energy via radiation and to finally collapse with the positively charged core, but it appeared to be possible to imagine the orbiting electrons as waves. Assuming that only full numbers of wavelengths can be sustained in a stable way (which depends on the rotating velocity of the electrons), quantization of electronic states is explainable. With the additional assumption, according to the Pauli exclusion principle, that only one or two electrons can occupy one orbit in form of a wave, the limitation of electron numbers in an atom is rationally understandable. We are dealing with the electron orbits first imagined by Niels Bohr. Today it is realized that the model is somewhat too simple and the orbits are less well defined due to the uncertainty principle. The latter essentially claims that due to the particle - wave existence of matter, a measurement process will always disturb a quantum system so that an uncertainty exists with respect to the measurement. The product of coordinate x and momentum p (velocity times mass) is always proportional to the Planck quantum of action h, meaning it cannot be measured arbitrarily accurate.

When a quantum system, for example an atom consisting of a positive core and negatively charged electrons, decreases its energy per state as the basic postulate claims, it will minimize its energy on the basis of relation (11). This also concerns orbiting electrons. Since energy oscillates between the particle and the wave state of electrons, then also the energy contained in information (E_n) has to be minimized. E_n describes the information needed for reconverting a wave into a particle. A simple standing wave will require a minimum amount of energy for information. A chaotic, unstable, or transitory wave will require a much larger amount of energy of information. The consequence will be that only states with

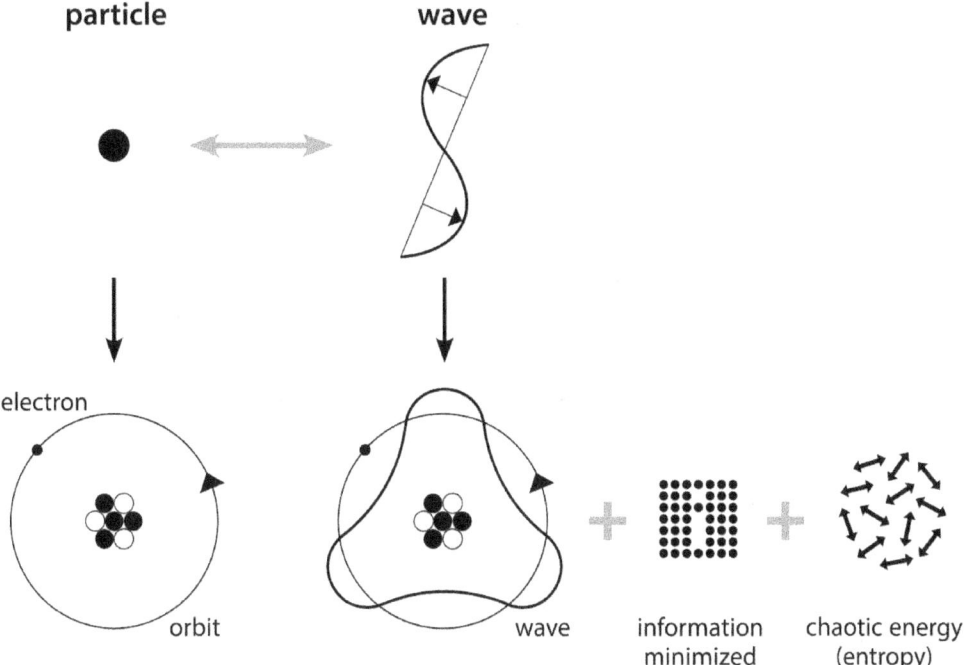

particle   **wave**

electron

orbit   wave   information   chaotic energy
minimized   (entropy)

Fig. 20. When an electron selects an orbit around a nucleus of an atom, it aims at minimizing its energy per state while also forming a wave. Considering relation (11) this means that also the energy needed for information has to be minimized. As a consequence only the simplest wave patterns around the atom will be selected. This is quantization.

well-defined stable electron density distribution (simple, standing wave patterns) will be selected for energetic transitions. This is the consequence of the system's search for a minimum energy per state and is consistent with experimental knowledge. It seems that this explains qualitatively the phenomenon of quantization. I consider this a very significant first result as it actually tells us why a system selects quantized energy values. If values in between were chosen, the information E_n for describing a much more complicated distribution of energy within the wave would be much larger. With an appropriate theory considering informational aspects, quantitative results may be expected. This may underline the reasonableness of the "symbolic form" identified for understanding quantum physics.

Going one step further one may also speculate why a free travelling particle can also appear as a wave with a periodically changing electric and magnetic field. Also in this case a most simple distributed form of energy, in terms of information needed to describe it, is a periodically changing travelling wave. Any more complicated energy distribution pattern would need more energy in the form of information (E_n). The drive or tendency to minimize the energy per state has led to a minimization of the energy relation (11) and thus to the wave pattern, because it represents a minimum energy contribution from E_n, the energy related to information.

Quantum correlation, the property which allows quantum particles to split up and maintain contact over apparently arbitrarily large distances, is a well-known phenomenon which has already been mentioned (fig. 7). It is a key for efforts to facilitate quantum teleportation and related innovative technologies (Zeilinger, 2007). Without going into greater detail it is claimed here that the information related to E_n, and available for the reconstruction of the particle state is involved in this phenomenon.

When particles are split up to yield quantum correlated systems, the information self-image existing for the original particle must also be split up and will have to become restructured to some extent. What does this information now do when the quantum correlated particles fly apart? It will depend on the information to be handled. There will be information which can easily be split up, and other information which maintains its relevance for both particles. To the latter, one may count the intrinsic angular momentum, the spin, of the system, which should be maintained for the entire quantum system. This is a well-known natural law. It can therefore be concluded that the splitting-up or doubling of information is only partial and some information will still have to maintain contact between separating and distant particles. An information link will therefore have to connect the separating particles. This is indeed a characteristic feature of quantum correlated particles. This would explain why manipulating one particle, for example by changing its spin, would have an effect on a distant quantum correlated second particle. Again a rational understanding of the phenomenon involved in quantum correlation is basically possible. The information link between two separating quantum correlated particles turned out not to be a "spooky" interaction. It was instead a necessary consequence of dividing up information, contained in the information self-image of matter of the original system. Joint information on the state of particles must be maintained, otherwise basic laws of physics would have to be violated. In relation to Bell's theorem it can be said that the phenomenon of quantum correlation can be rationally understood, but information on matter and energy has to be included into the description of quantum reality. And part of this information is responsible for the observed long distance interaction. Compared with a classical picture it would be as if two people in a city separated and maintained communication via cellular phone. They are searching in the shops for two different articles. When the first finds one of them in a shop, the second receives the information and concentrates on finding the second. I still consider this to be part of local classical realism. It, however, became only possible after the cellular phone was invented. The cellular phone works on the basis of natural laws and nature also applies its laws and is able to use information and telecommunication. As with the real world it cannot be claimed that it is non-local. Communication at a distance may occur only in special cases when joint information on separating particles cannot easily be subdivided. Later it will be shown that the phenomenon of gravitation is involved in this quantum correlation.

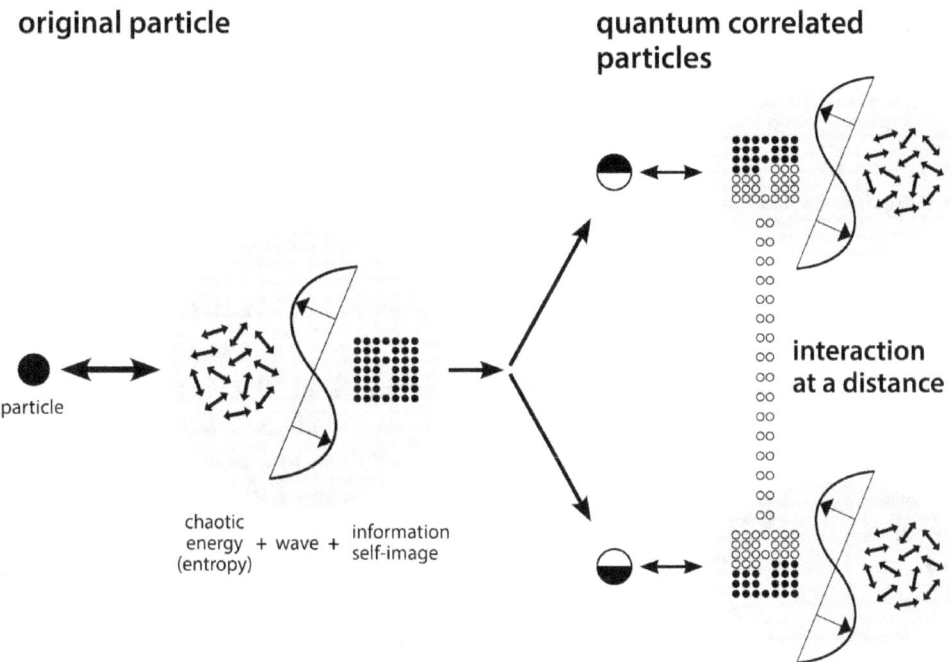

original particle

quantum correlated particles

particle

chaotic energy (entropy) + wave + information self-image

interaction at a distance

Fig. 21. When quantum correlated particles split up, the information on their particle-wave state will also have to split up. But some common information structure must be maintained if required by conservation laws. It can be imagined that it is this information which keeps the mysterious, "spooky" correlation contact.

Information on the particles involved would thus be a key to a deeper understanding of quantum correlation. Reality and information about reality would accordingly be involved in a mixed function. Only if both are considered the quantum phenomenon is rationally understood. As visualized in figure 21 a trace of "joint" information links the quantum correlated particles. This does not mean that information itself travels between the two correlated particles. It just links them and upon measurement on one correlated particle, the remaining complementary information will be implemented on the second. Such a spatial breakdown of joint information may not be limited by light velocity, since one does not deal with information transfer by light or matter, which appears to agree with experimental observation. Later, in chapter 21, it will be seen that the self-image of information can be identified with gravitation. Gravitation needs mass to occur, but it is not a consequence of action. Since action is the consequence of a cause which requires time, gravitation should not be exposed to the flow of time. Hence, a process occurring within a "gravitation link", as shown in fig. 21, could be instantaneous.

Such an information-based picture of elementary processes can also give a more profound access to the secrets of quantum physics involved. The information concerned will have to be studied on the basis of a more detailed theory, which still needs to be developed, and again it can be hoped that a deeper understanding can be reached.

Concluding these considerations, it is possible to say that quantization and quantum correlation as well as the particle-wave duality in its extended form, as described by relation (11), directly follow from our initial postulate that energy aims at decreasing its presence per state.

14 What does quantum uncertainty mean in nature?

There is another important quantum phenomenon which has changed our way of recognizing and understanding nature since it is now accepted as fundamental. It is described by the uncertainty relation first proposed by Werner Heisenberg in 1926 and attributed to the parallel wave properties of a particle. The uncertainty relation states that position and momentum as well as energy and time of a quantum object cannot be measured exactly, but only with an uncertainty which is proportional to Planck`s constant of action (h). This became a mysterious basis of quantum theory which thus adopted an element of unpredictability and uncertainty. Not the reality of particles is described in quantum mechanics, but our statistical observation and measurement of these particles. Einstein was not amused by such a perspective and is known to have commented: "God does not play dice".

The claim of Heisenberg`s uncertainty relation is, as Karl Popper (1979) already realized, in contradiction to both the a priori causality principle of Immanuel Kant (1958) and Ludwig Wittgenstein`s (1918) thesis on the omnipotence of science: when the position of an electron is known, its velocity cannot accurately be determined.

How can a game with dice be understood within the "dynamic energy" hypothesis? Does this mathematically deduced quantum phenomenon really mean a departure from causality and rationality? In equation (11) the frequency v (one over time) can be transferred to the right sides as t (time), and the result then multiplied by E_p, the total energy of the particle, in the numerator and denominator. This way an equation is obtained, which has the structure of the uncertainty relation: $E = E_p$, the total energy of a particle, multiplied by t, the time, is proportional to h, the Planck constant of action. What is interesting now is the proportionality constant which shows up. It is the ratio of the energy of the particle and the energy of the wave (made up here of distributed energy, entropic energy, energy of information). The magnitude of the proportionality constant will thus simply vary and depend on whether the energy is more present in the particle or in the wave at the instant of measurement . This suggests that the uncertainty is related to the intermediate presence of energy in form of a wave, and specifically, in the form of information on energy for the reconstruction of the particle from the wave. It may be, depending on the snapshot of inter-conversion between particle and wave captured during the measurement act and related to the presence of information, that fluctuating and partial particle properties are reconstructed from a wave. This result clearly indicates that the phenomenon of uncertainty is related to the wave nature of particles. Here, also the information self-image of the particle enters, but there is no reason at all to assume a fundamental origin of uncertainty. It is true that a detailed theory for our interpretation of the particle -wave dualism still has to be developed. A deterministic statistical chaotic mechanism as a reason for the observed uncertainty is definitively to be expected. The proposed process of inter-conversion between particle and wave is responsible for a statistical reconstruction of the particle's properties. There is no reason to assume a deviation from causality. The uncertainty phenomenon is merely shaped by a complex natural phenomenon including information turnover. With its complexity it is veiling a logical physics behind.

It is well known that the uncertainty relation also predicts that a "zero-point" energy remains when all other energy is removed from a system. If this were not the case, position and momentum of a particle in its lowest energy state would be known at the same time. The existence of such zero-point energy is well established in the science community. A sea of such energy should however be spread all over the universe and such energy should pop out from nothing. An absurdly large amount of zero-point energy results with a discrepancy between theory and observation of 120 orders of magnitude. This mysterious energy is in the centre of strange speculations on free energy harvesting from space and on dark energy. I conclude that such a phenomenon is not real, but feigned by the discussed wave- and information aspect of the particle-wave duality. When particles are in a real minimum position, they should behave like that. But popping out of energy from nothing is definitely impossible on the basis of the new understanding of the Heisenberg uncertainty relation.

Later it will be shown that elementary particles can be imagined as products of self-organization of energy. It is therefore probable that a deterministically chaotic process is involved in quantum uncertainty. As a consequence, God is not playing dice, but we observers are ignorant, faced with a mechanism during which small parameter changes may generate large statistical consequences during measurement.
Since such a conclusion is very important, let us again look at the situation with the uncertainty condition. The uncertainty in quantum physics appears to be

relevant when the wave is interpreted in relation to the particle. Waves alone can well be described in a deterministic way and this is done using the so-called Schrödinger equation. The outcome is nevertheless probabilistic, but nobody seems to understand exactly why. In my interpretation of the particle - wave duality (11) it is thus the back-conversion of the wave (right side of relation (11)) into the particle, which generates uncertainty during a measurement process. There is obviously an interference with information contained in E_n.

The outcome of a quantum measurement apparently depends on the state and handling of information E_n, which when only transferred partially, may bring chaotic attributes.

Such a conclusion about uncertainty being the consequence of a chaotic mechanism of recovery from a measurement, has quite dramatic consequences. The world is not fundamentally uncertain and non-deterministic, as theorists of quantum mechanics claim. Only chaotic measurement conditions are involved with deterministic statistical chaotic results. With adequate experimental techniques the outcome of such measurements may be improved. I recall that when I studied physics during the sixties of the past century, I learned that due to the uncertainty relation, one will never be able to see an atom. Today, using tunnelling electron microscopy one can very well observe the beautifully ordered patterns in which atoms are arranged in interfaces of solids (fig. 22). One can even visualize how the atoms move around and interact with each other.

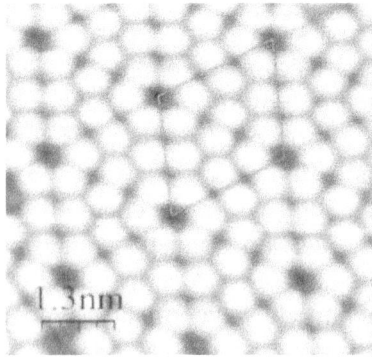

1.3nm

Fig. 22 Modern tunnel-electron microscopic images (here of a silicon surface) allow the visualization of the arrangement of individual atoms on surfaces. In the past, more stringent interpretations of the uncertainty relation did not permit to expect that. (credit: B. Kaiser, Technical University Darmstadt, the scale is 1.3 billionth of a meter).

After identifying the uncertainty problem as a consequence of measurement interference during the back-conversion of the wave into a particle via information, one can draw an important conclusion: within the discussed dynamic energy model, there is no basis for claiming that particles, and thus energy, can pop up from nothing. Such a mechanism is, for example, invoked in the inflation theory of space after the Big Bang to explain the generation of the vast amount of matter observed in the universe. It is also used to derive the Hawking radiation from black holes, or to explain riddles in the microwave background radiation on the way they are expected to take towards the formation of galaxies. Without the possibility, via the uncertainty relation, to generate matter and energy from nothing, relevant presently favoured space theories would simply collapse. Most importantly, there would be no basis left for claiming non-causality. Such effects without reason have pestered critical spirits for one century already as they contradicted human experience, but quantum theory appeared to make them unavoidable. The Casimir effect, an attraction of two metal plates across a very short distance, is usually mentioned as a phenomenon supporting this quantum interpretation of particles popping up from nothing. However, it can easily be explained classically on the basis of weak (van der Waals) interactions between the plates. It is the interaction between polarizable surface states, sites with displaceable electrical charges, or surface molecules (Jaffe, 2005).

15 Paradoxes can be eliminated

As another example and to further demonstrate that the dynamic energy approach makes sense, I will now explain how the much discussed double-slit experiment becomes intuitively understandable. The experimental set up of a double-slit experiment is very simple (fig. 23). Photons or elementary particles are allowed to pass through double-slit openings in a screen, behind which a diffraction pattern is observed. It is quite characteristic and different from a pattern observed with just one slit, which is quite uniform. This double-slit diffraction can easily be understood, when the wave properties of light are taken into consideration since the length of the light path varies with the angle, leading to a superposition and annihilation of waves which generates the interference pattern. The problem here is that while the wave concept can explain the observed diffraction pattern behind a double slit, the particle concept cannot. Strangely, the experiment also works when very low illumination density is used and light particles are expected to move through the slits one by one. When elementary particles, such as electrons, which have a wavelength related to their energy are passed through the slit, exactly the same diffraction pattern is found, and it is also found when the electrons pass through the slit one by one. How can individual particles passing successively through one slit know of the second slit to generate the identical diffraction pattern as a wave? The quantum theoretical explanation appears to be obvious, but is irrational: a passing particle seems to be present simultaneously in both slits. It assumes two different positions.

The proposed concept of a dynamic energy which both facilitates an oscillation between particle and wave and is able to activate energy for information, E_n, for the compensation of entropic energy (due to the expansion of a particle into a wave (11)), can give a straightforward intuitive explanation (figure 23). The information which appears with the wave pattern will recognise and image the geometric form of the double slit (as waves can). This information, activated during the wave state, will be recovered for the particle during re-conversion of the information energy (negentropic energy). In this way, individual particles successively passing through one slit will also recognise the second slit and may generate the diffraction pattern characteristic for the corresponding wave. It is the energy in the form of information, E_n, activated to image the expansion of the wave (around the double slit obstacle), which is responsible. The double-slit experiment can thus, by consideration of information as part of the particle-wave duality (relation (11)), be understood in an entirely intuitive way. The information collected by the expanding wave within the double slit geometry has a back effect on the fate of the particle, when it leaves the slit. The new "symbolic form" which I propose works. The rationally thinking brain is satisfied with the given explanation.

It should be recalled that classical moving particles select a path following the principle of least action (fig. 16). For quantum systems, in contrast it was found that they seem to consider all possible paths towards the final destination. In the case of the double-slit experiment (fig. 23) an individual particle would select paths through both slits. Such a behaviour, a double path for one particle, is irrational, but can be calculated and indeed explains the observed patterns produced on the screen. The calculation technique applied, called the sum-over-paths method, contributed to the Richard Feynman's Nobel prize. How would I understand such a sum-over-path phenomenon in the light of the proposed alternative interpretation? A particle would start following the principle of least action. Then it would change into a wave, equivalent to many distributed tiny energy dots. Each of these would continue following the principle of least action and cross through one of both slits. Behind this, the tiny energy dots would regroup to form a particle again – a particle which has "seen" both slits. The phenomenon remains rationally understandable.

What evidence exists that photon energy really changes and oscillates between the particle and the wave state? This can be derived from a simple consideration. A photon approaching an interface with light velocity does not know whether it will encounter light absorbing molecules which interact with a particle, or a light diffracting structure which interacts with a wave. Nevertheless, both mechanisms can immediately be activated. Both the particle and wave state must consequently be active within a very short time interval.

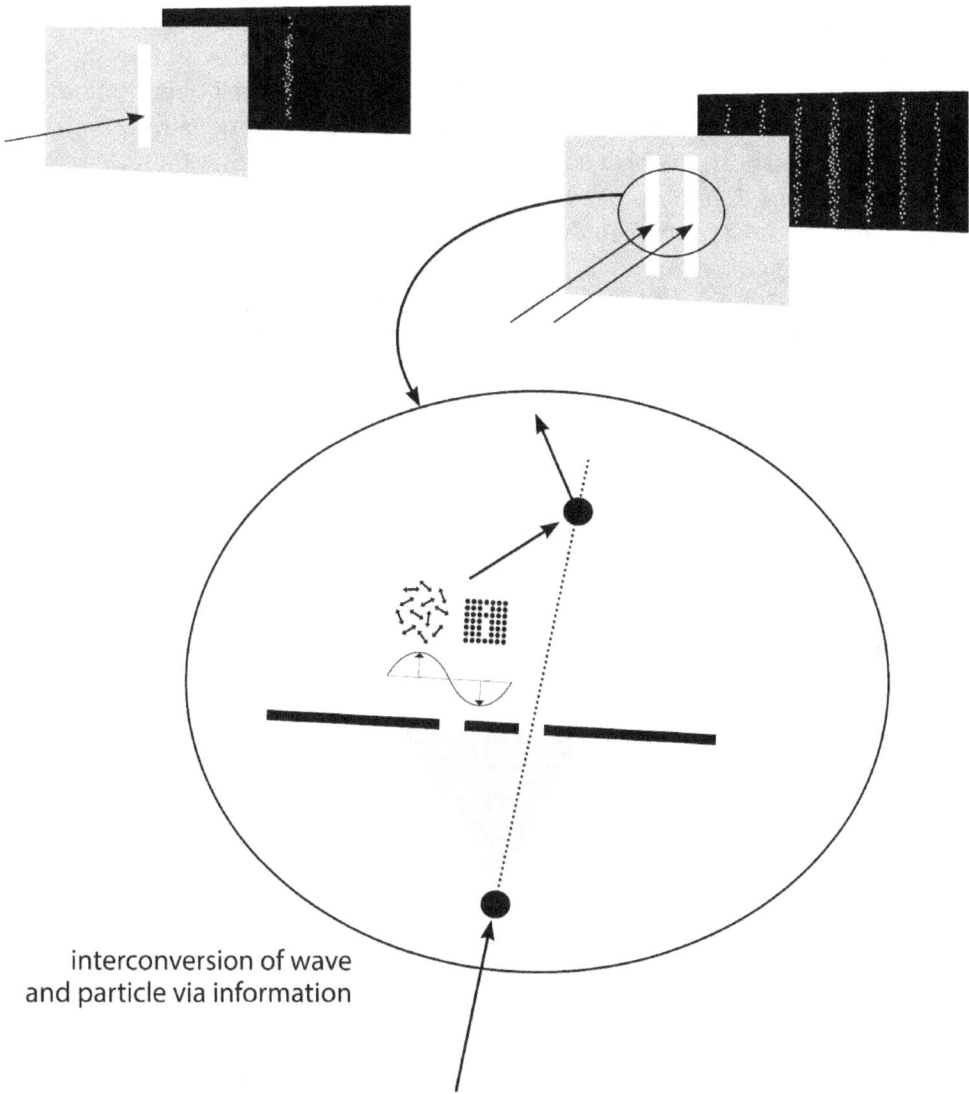

Fig. 23. Diffraction patterns for light and elementary particles are different behind a single and a double slit. Even when particles pass individually the patterns do not change. On the basis of the new interpretation, they see the second slit due to the proposed particle wave duality controlled by the information self-image.

16 What can be said about non-locality in general?

Non-locality, or action at a distance, is, as we have seen, an extremely intriguing property claimed for quantum physics. It is derived from the wave property of quantum objects and describes their ability to communicate instantaneously over arbitrarily large distances if they are quantum mechanically linked or entangled. A rational explanation of entanglement was given above (fig. 21) using a partially split-up information self-image of matter. The phenomenon can also be deduced from the double-slit experiment. It was traditionally interpreted in such a way that one particle can be simultaneously be present at two clearly separate positions. A particle passing through one slit knows about the other slit. Consequently it must have been there.

What is the supposed origin of non-locality in quantum theory? A wave is distributed over a much wider range of space than a particle. If one, as is done in traditional quantum physics, interprets the distribution of the wave as the probability of finding the particle in a specific location, then the particle can be in very different locations. If the particle is now measured at one location, then the probability of finding it at a specific distant location collapses. The particle can no longer be elsewhere. This collapse is assumed to be instantaneous. The local measurement thus has a non-local effect on all distant locations where the wave function is still present. The phenomenon is especially remarkable when two or several particles are correlated and linked. Then immediate action at an arbitrary distance seems possible because the measurement at one location will affect the probability of existence of a particle in an arbitrarily distant location. When this phenomenon was first discussed, Albert Einstein dismissed it as "spooky action at a distance" (Einstein et. al., 1935). But it was later pinned down by the famous paper by John Bell in 1964 (Bell, 1964) and with experiments by John Clauser and Stuart Freedman in 1972, and by Alain Aspect in 1982.

What is now different with the dynamic energy approach discussed here? An important fact is that the spatial distribution of the wave does not reflect the probability of finding the particle. The wave does not describe the localization probability of the particle, but it is the consequence of the particle's effort to decrease the energy's presence per state via distributing the energy into space. As explained before, and described in the form of equation (11), distributed energy, non-usable "entropic" energy, and energy in the form of information is involved in the "wave" pattern (fig.19b). Since all these together should reflect a minimum quantity of energy per state within the constraints of the system, the information involved is also minimized. Definitively, one cannot associate the distant presence of wave energy with a probability to locate the particle there. This means one cannot speak of the non-locality of a particle or of an object. Instead one is dealing with a quantum system, which is trying to decrease its presence per state by expanding into space. Such an irrational phenomenon as non-locality appears not to exist. This could already be found when interpreting the double-slit experiment (fig. 23). Also this non-locality paradox is thus eliminated via the symbolic form of a "dynamic" energy. However, I should repeat that this is my personal opinion and not the opinion of the quantum physics community today. It relies on experimental quantum behaviour as observed by Clauser, Freedman (1972) or Aspect (1981). The latter researcher, in experiments conducted with J. Dalibard and G. Roger (1982) on quantum correlated photons, appears to have demonstrated that he was able to falsify Einstein`s concept of hidden variables. No addition of variables to the theory can improve it. The consequence is either "spooky action at a distance" or the verification of the non-locality interpretation of the Kopenhagen school. The conclusion given is that quantum non-locality is a "property of the universe" that is independent of our description of nature" (Quantum nonlocality, 2014). This is a very arrogant statement. It is equivalent to saying that an important foundation of nature is fundamentally irrational to our reason. Is this really true? On the basis of the arguments presented here it is impossible to accept such a conclusion. Our "symbolic form" analysis suggests a different situation. There is no need to postulate non-locality. And the "proof" must be wrong or is misinterpreted. It is the information on energy E_n, involved in our dynamic energy concept (relation (11)), which is obviously responsible for the observed quantum behaviour. It was shown above (fig. 21) how this self-image of information already resulted in an understandable concept of quantum correlation, the phenomenon which links particles at a distance. This is obviously a real phenomenon which could be explained via an understandable property of the self-image of matter. The "spooky action at a distance" is not spooky, but real and, in principle, understandable. If information is important for both separating particles, because a natural law prescribes its conservation, it continues to link them. No other theory up to now could explain this on a rational basis. A better elaborated theoretical concept is, of course, nevertheless needed to understand and describe more fundamental details. Sometime in the future such a theory should be available, but a non-locality property of particles will not be the finding. The non-locality paradox is, in my opinion, off track. It is the consequence of a poor description of physical reality. We are dealing with an information phenomenon which is responsible for the observed experimental behaviour. It is a property of the information self-image associated with the particle-wave duality of matter which we still have to learn to better understand.

Another counter-intuitive phenomenon of quantum mechanics is the tunnel phenomenon. It describes the finding that an atomic particle can penetrate a potential barrier even when its energy is smaller than the height of the barrier. The radioactive decay of atoms via the emission of alpha particles offers one example. Based on the new interpretation of the particle wave dualism (11) with information as a mediating form of energy, the explanation is quite simple. While interacting with the barrier, the particle transforms into a wave with its contribution to entropic, chaotic energy, and the information which may restore the particle again. It would partially penetrate the barrier and information may, with a certain probability, restore the particle again on the opposite side. The new "symbolic form", a particle-wave dualism including information, helps our intuition to understand the tunnel process rationally. Knowing from our present experience that a 3D printer can indeed transform information into a three-dimensional object gives us the confidence that we are dealing with a logically explainable phenomenon. Recently it was communicated on television that an entire drivable car was printed via a 3D printer with plastic material within two days. There is now industrial interest in applying this technology for the production of a wide range of commercial products. They comprise sophisticated spare parts and household articles as well as food products and toys. One company is scanning people and producing a miniature version with perfect shape and colours. Reassembling objects using pure information is definitively possible! Later, (in chapter 21), it will become clear as to why the information self-image, which reassembles the particle, can easily penetrate energy-matter barriers.

As a further quantum paradox one should address the thought experiment of Schrödinger's cat. Erwin Schrödinger intended to use it to demonstrate the bizarre properties of quantum theory and its possible incompleteness. A cat is locked into a box and exposed to a quantum process which can trigger the release of poison. Quantum theory allows the quantum systems to function simultaneously in superposed states, active and non-active. The wave function of Schrödinger's equation describes that, at least its squared absolute value, which provides only statistical information. Until the box is finally opened to find out the state of the cat, the animal should also be in the combined state of dead and alive. However, this is not credible on the basis of human experience. It has been emphasized before that in the new interpretation discussed here the strange statistical outcome of quantum states is due to the information, which mediates between the particle and wave state (fig. 19b). It is the need to add to a description of reality, which is particle or wave, the required information on this reality. This is logically understandable. The superposition of a "dead and alive cat" is the superposition of particle and wave as long as no energy is being converted. When energy is being converted and the box is opened, the system decides between particle or wave property with deterministic statistical probability, because energy mechanisms follow self-organized pathways.

What if the here given explanations are basically correct? What should we then think of the mathematical and experimental "proof" of non-locality and irrationality on which the science community has been insisting? Would it not have been more reasonable to admit that science does not yet understand and is still searching for a credible rational mechanism?

RETURNING TO A CREDIBLE COSMOLOGY

17 The ever constant light velocity: what does it really mean?

A relevant newly proposed concept for quantum physics should also have implications in related science areas. We shall therefore concentrate on the theory of relativity, which is difficult to match with present concepts of quantum physics. The measured constancy of light velocity irrespective of the relative movement of the detector going towards or away from the light source is a paradox, which cannot be understood intuitively by the human brain. The crucial experiments that demonstrated an ever constant light velocity started in 1887 with Michelson and Morley. They measured the velocity of light arriving on the earth from space going with and against the direction of earth movement. They found no difference. This experimental fact is very puzzling and incomprehensible to a naive observer. He is accustomed to seeing an effect of a moving object, when a signal is emitted from it. An example is the sound heard from a train which is passing by. It is different from an arriving train and from a departing train, because the velocity of the train is added or subtracted from the velocity of the sound. Car drivers also know that a crash is much more devastating when the cars involved come from opposite directions (fig. 24, bottom) compared to a pile-up collision.

An ever constant light velocity is therefore just accepted to be an irrational fact. Apart from a later rejected "ether" theory it has found no explanation until it became a basis of Einstein`s theory of relativity. In it a ghostly space had to be invoked with the incredible property of always keeping the light velocity constant for an observer, regardless at what velocity he approaches or recedes from a light source. The irrationality involved is the more obvious as the empty space around the travelling light particle must somehow realize at which light velocity the observer is approaching in order to immediately adjust the observed speed of light for the reference system. In addition, manipulation of space itself and of time occurs when approaching mass is involved. This is considered to occur automatically via the empty space in the general theory of relativity. And this should happen on the basis of time-invertible basic laws which do not care about time orientation? Nothing we know about physical mechanisms can help us to understand.

Before Einstein`s theory of relativity, significant efforts were undertaken to explain the always constant light velocity with the concept of an ether- a very light, fluid substance - which was expected to mediate the propagation of light. Poincaré and Lorentz applied the relativity principle while imposing a constant light velocity. They ended up with the famous Lorentz transformation, which later also Einstein used. Its real pioneer, however, was W. Voigt in 1887. Dilation of time and contraction of objects were calculated and assumed to be the consequence of an ether, which imposed the constant light velocity. Einstein concluded that consideration of an ether was not necessary for a relativity theory and simply assumed that empty space assured that light velocity is always constant. The basic postulate was that every relatively moving object has to be subject to the same natural laws. And an ever constant light velocity was considered to be one of them. Of course, space thus turned out to be very special. It became four-dimensional, with a fourth dimension controlled by distances determined by light velocity and time (multiplied they give the fourth dimension of length).

The irrational fact of an always constant light velocity has thus become a "symbolic form" for the theory of relativity, which itself, however, is not accessible to rational cognition either. Our brain does not allow us to imagine our space with a fourth dimension, but mathematically it can be defined that a point is not only determined by three coordinates of position, but additionally by the distance light can travel during a certain time and with constant velocity for every observer. The consequence is a four -dimensional space which is manipulated and bent by masses to explain gravitation and inertia as well as constant light velocity.

Can the new concept of the particle–wave duality of light discussed here provide relevant innovative insight? The outcome, in fact, is very simple: when assuming a dynamic particle-wave inter-conversion as expressed in relation (11), it is clear that photons may be continuously reassembled from energy in the form of information (negentropic energy E_n). This means that the information self-image continuously reassembles light particles from light waves (fig. 19b).

What does it now mean when light arrives in the form of information E_n and distributed energy and reassembles to generate a photon particle E_p? It can be imagined that photonic energy is transmitted and photons are reassembled in a similar way to show light particles with constant light velocity as television images are reassembled from transmitted digital information. The latter process indeed functions entirely independently of the relative velocity of the television receiver (e.g. on an airplane). Independent of the relative velocity, a sharp picture is generated from information in the form of digital signals. In a similar way, because of the involvement of energy in the form of information E_n (related to negentropy), the arrival of photon energy in the direction of and contrary to the receiver velocity would also not matter within the presented concept. The result would always be a light particle with the given identical light velocity (in the velocity frame, where the measurement takes place). There would be no need to bend and deform space to always fix a constant light intensity.

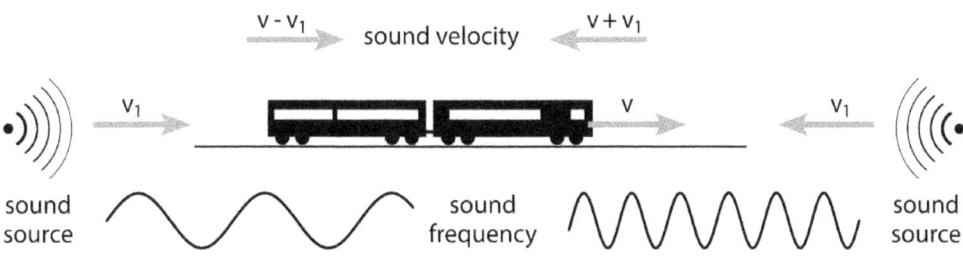

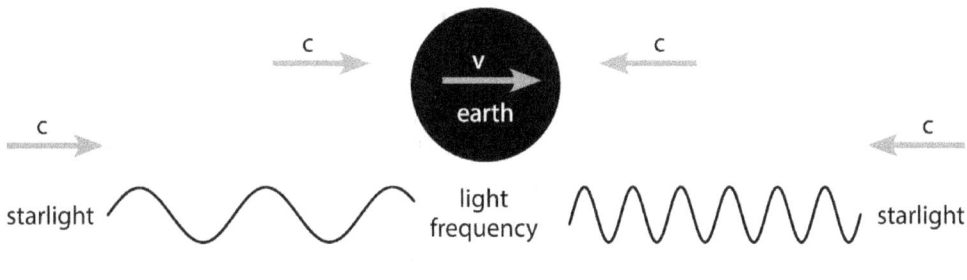

Fig. 24. While the velocity of moving objects (bottom), or of sound (top) adds or subtracts, light does not care about relative velocities (centre). Its velocity in free space is always measured to be constant. The theory of relativity claims that this is due to elaborate properties of space itself. In contrast, the dynamic energy model discussed here recognizes it as an intrinsic property of light, which continuously reassembles the particle from a wave via an information self-image.

It can intuitively be understood that different velocities of the light registering systems would not matter. How could such a velocity of a moving system have been detected by space around an arriving photon anyway? In contrast: independent of the velocity of the light-measuring system, information would be converted into a photon always with the same maximum light velocity, because this would be determined by natural law. The information self-image of matter "knows" that an imposed light velocity cannot be trespassed. Not empty space but the photon's information self-image itself does the job and no additional assumption is needed beyond the modified particle-wave duality.

We are presently living in a period with a boom in information technologies. Cell-phones, Internet and computer-driven machines have changed our world. And it seems just to be the beginning. Where will our information technology be in a hundred years? Information systems simply exploit physical laws. They reflect what nature is able to provide in terms of information technology when appropriate conditions are given. Three-dimensional printing, for example, allows the production of real objects starting from pure information. It is exactly something like this which is claimed to happen here when a photon is reassembled from a wave. This is not irrationality, but rationality which allows us to understand what happens with the photon during measurement. Three dimensional printers actually work and one can understand how they work. They can, in principle, create a toy car which is able to propagate at a given speed within a flying airplane (fig. 25). And the same speed of the toy car could conveniently be adjusted in airplanes which are flying in the opposite direction or with widely varying speeds. With the knowledge that information can be used to reconstruct an object with given properties, the "symbolic form", the platform which shapes our perception, helps us to understand. The reconstructed photon can be expected to propagate at the velocity characteristic for it, the velocity of light. There is no need to invoke a mysterious four-dimensional space in order to instantaneously simulate a constancy of light velocity and to manipulate time. Information, and all its astonishing properties, has been recognized to be active in a basic law. From relation (11), which links a particle with a wave, it can be deduced that the reality of a particle and a wave is coupled with an information self-image to reconstruct a particle and its properties from a wave. Reality and information on this reality are linked. Both are needed to understand quantum processes. And the information on photon properties is the reason for a constant light velocity being able to be imposed. It implements a fundamental law.

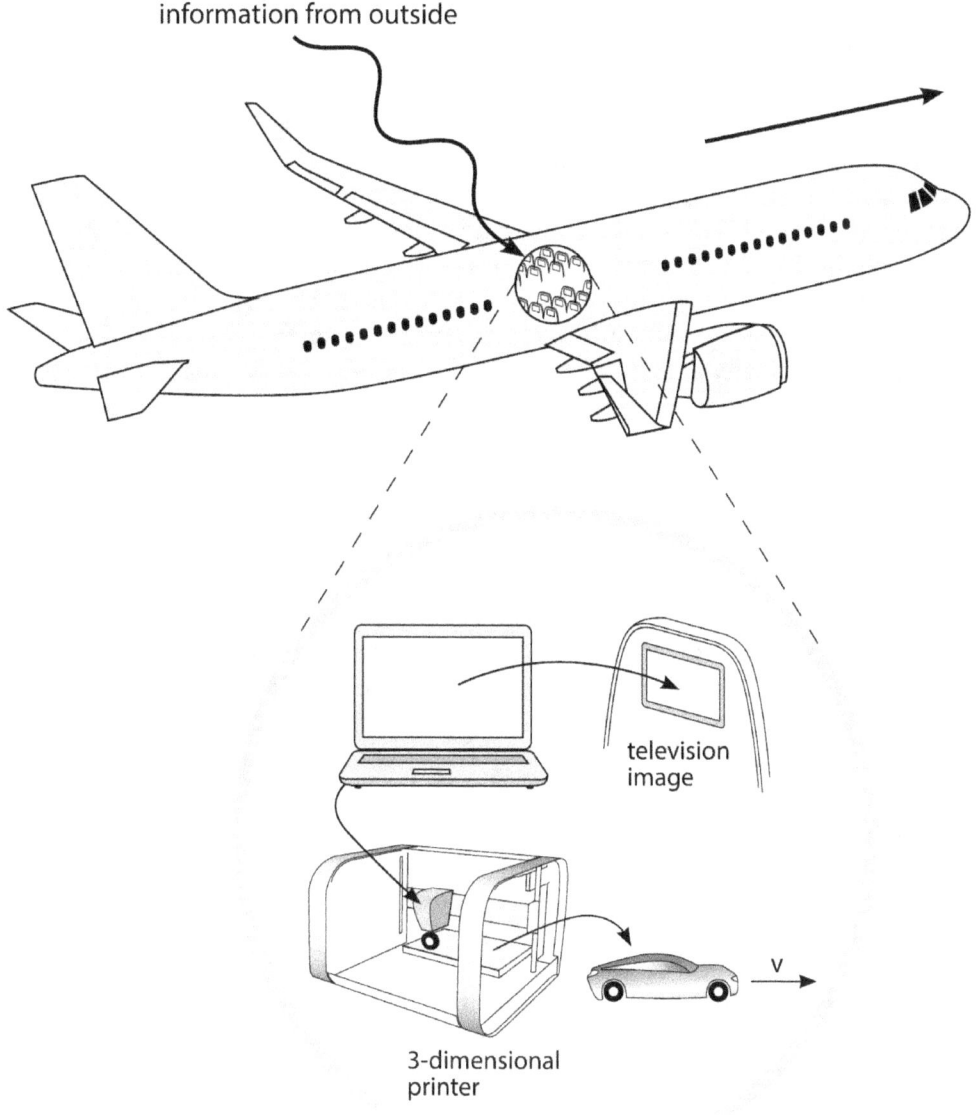

Fig. 25: Why can information reconstruct light particles with an always identical velocity? It is like transmitting a television program or the information for a three- dimensional printer to a flying airplane. The printer could produce a toy car which drives with a given velocity. The outcome is fixed and will be identical irrespective of the airplane's flight direction and speed.

The paradox of an always constant light velocity could therefore be understood entirely intuitively. A photon would be reassembled the same way according to relation (11) to yield the given light velocity in the system where the measurement is taking place. The mystery of an ever constant light velocity independent of the velocity of the light source or detection system would simply dissolve. Since light would be converted via intermediate information, different relative velocities would be irrelevant and we would intuitively understand what is going on: an information self-image of the light particle controls and adjusts the natural phenomenon of constant light velocity. The "symbolic form", the constancy of light velocity, which became the fundamental law and which was used to shape the theory of relativity via a manipulation of space, would in this case become rationally accessible and understandable. No theory was able to do that until now, which suggests that the dynamic energy approach to the irrationality of quantum physics makes sense. It can be extended to re-evaluate the theory of relativity and, hopefully, additional theories that create irrationality. The rational world created is simpler and understandable.

However, while a superficial cognition has been attained by the notion that information can be used to tailor constant light velocity, a detailed physical theory about the handling and function of information in quantum physics still has to be developed. This appears to be a significant research challenge for the future. This effort would be worthwhile, because the alternative is a puzzling four-dimensional space-time, which imprints a constant light velocity into all natural situations. Space and time interact with each other and adjust themselves in such a way as to always enforce a constant light velocity. This is a tremendously complex task and cannot be logically understood. Neither the four-dimensional space is conceivable, nor is a completely empty space with a property to always force a photon into constant velocity when a measurement occurs. What intricate mechanisms could be active in an empty space to develop such flexible and demanding properties?

There are also additional consequences from the explanation of constant light velocity presented here: new fundamental symbolic forms (fundamental laws) now apply within the presented approach and will also give new critical insight into the statements of the theory of relativity. It should be remembered that up to now quantum theory and the theory of relativity could not be unified. The search for a quantization of space and time has equally not been successful. Now we can suspect why there were complications. One has to rethink both quantum theory and the theory of relativity. The first is incomplete and affected by an error. The second is superfluous, because constant light velocity is not a property of space. In part, this rethinking will be done in the following chapters.

Firstly, I would like to briefly recall my personal experience when reflecting upon the meaning of the self-image of information, which in my energy concept means handling the particle-wave dualism. It was in January 2014 and I was skiing in a ski-resort near my home as I do frequently in wintertime. It was a weekday and I was nearly alone on the slopes. I enjoyed this solitude and tried to understand where this information self-image of matter, which I was convinced interferes with matter, could have a detectable impact on nature. For some time I had no relevant idea. Then I realized that with the information from my brain I could comfortably control the velocity of skiing. Why should not the information, associated with a photon, adjust an ever constant limiting velocity of a photon during a measurement? And why could this velocity not be adjusted when the medium changes through which the light propagates. It would be like changing from a well-prepared skiing slope into deep, untouched snow. The consequence of these considerations was initially shocking for me: an irrational space-time would no longer be needed as an explanation. I remember when I got this idea, I had to interrupt the downhill trip, because this idea distracted me too much. Later I mentally searched for an observable phenomenon which is characteristic and representative for the information self-image of matter. It must somehow create an energy halo around matter. What is it? There was finally no alternative left besides gravitation, a well-known phenomenon, which is however still only poorly understood today (see chapters 18, and 21). Gravitation is associated with matter and behaves exactly as an information self-image should behave. Nature is rational and to understand its mechanisms, it is worth being patient. It was an enjoyable day and the cold, fresh air in the mountains was apparently favourable for reflecting on rationality in natural laws.

18   The four-dimensional space-time: an illusion

After learning from the dynamic energy approach that the experimentally well established constant light velocity could be understood as a local phenomenon of light particles and not as property of the whole of space, one has to look at consequences. Einstein`s relativity theories, both the special and general one, which adopted the constant light velocity as a property of empty space and led to the concept of a four-dimensional space-time, are nowadays entirely accepted in the science community. Space and time have become dynamic quantities. Moving objects affect space and time around them and space-time forces objects into well-defined orbits. Such a theory is highly irrational, firstly because one is dealing with the void, which should contain nothing, secondly because it gets an additional dimension which nobody can imagine, and thirdly, because this void is given fantastic properties. It can even manipulate time around matter, a capacity which nobody can imagine, and which cannot even be thought of being possible using the most complicated machines. In the past such an ability of manipulating timeflow was attributed only to gods and some of their messengers, the shamans. This is not surprising since changing the timeflow, according to my understanding, means that naturally controlled constants such as the gravitation constant or rate constants, light velocity, or life-times of excited molecules have to be manipulated. How can this happen at all and how should empty space do that regularly and dynamically in response to matter in its neighbourhood? In addition, the time Einstein uses is just clock time, an ordering parameter used to monitor change. How can such a simple theoretical parameter, just because one applies it to form an irrational four-dimensional space-time, control and describe reality in such a complicated manner?

But these hidden properties of the general relativity theory are tacitly accepted, because nobody seems to have an idea on how to explain the experimentally verified always constant light velocity using a different mechanism other than by a space with all these assumed, very mysterious properties. But there is, of course, also a long story of criticism of the theory of relativity documented in the Internet (conservapedia, 2014, relativity criticism, 2014), and there are also arguments for its refutation. In part, the controversy about the two theories of relativity has been quite emotional with critics accusing the science establishment of avoiding open critical discussions.

The dynamic energy approach presented here opens the opportunity to explain the always constant light velocity in a different and very elegant straightforward way, and not via complex properties of space, which could be entirely avoided. This justifies a critical reinvestigation of the thus changed situation. The proposed modified particle- wave duality, a straightforward consequence of the "dynamic energy" postulate, can impose a constant light velocity via the involved information self-image of matter. No additional assumptions are required. The need to re-establish a particle from a wave via information simultaneously provided the possibility to generate the always constant light velocity during a measurement procedure or a photon energy conversion. It just has to be accepted that nature implements an always constant, maximum light velocity as a natural law. This is actually what is observed in experiments. This finding of an alternative to a four-dimensional space-time, implementing a constant light velocity, opens the potential to eliminate dramatically irrational concepts in physics. Is this really an option after hundred years of verification of the relativity theories?

To better understand what advantage such a new alternative for the explanation of an always constant light velocity brings, let us briefly investigate further what the special and general theory of relativity actually mean. During the first decades of the past century, after the theory of relativity was proposed by Einstein, there was significant criticism from established scientists. Now it is not only supported by the science community, but also widely popularized, discussed on television and even instructed in schools.

Free space was originally expected to behave entirely homogeneously, that is to be entirely empty and devoid of any internal structure and peculiar properties. It was used as an absolute reference, as Isaac Newton saw it. The theory of relativity then gave space very special qualities. The space became four-dimensional, giving to light velocity a decisive role in shaping reality, in manipulating geometrical structures and time in dependence of matter. Space became curved and the tracks, the trajectories of light around matter became distorted. This way, and also because time is "bent" or changes its flow rate, objects travelling through space-time also experience an acceleration. This is understood as gravity. Time, when subject to gravitation, will proceed in a special way around matter.

Closer to matter, time will proceed more slowly and when distant from matter, it is expected to proceed more quickly. Sending a signal towards a mass could thus in principle even invert time. Time could flow backwards. Has time in the theory of relativity thus not been abolished, when it ceases to proceed uniformly? It looks like actually being the case. But in spite of all these complications within the general theory of relativity, one cannot get rid of time. When travelling through space, one unwittingly also travels through time, as Einstein defines it "the time a clock shows". As mentioned, a four-dimensional space is a space which needs four parameters, or quantities, to specify one point. For space-time in addition to three space coordinates, a time dependent coordinate is needed. It is determined by the speed of light multiplied by time. This time-like coordinate determines the so-called event horizon, because light velocity limits the possibility of causal events and imposes itself as an ever constant property of light. Over the years a number of experiments have been conducted in support of the theory of relativity. For example, very accurate atomic clocks were placed on commercial airliners on trips around the world. An expected time shift of fractions of a microsecond, measured against a clock left behind, was indeed observed. More significant effects are found with elementary particles, muons, created by cosmic rays in the upper atmosphere. They apparently cover the distance (l) to the earth's surface much faster then they should on the basis of their velocity, as if this distance had contracted. Such a contraction is actually expected in the theory of relativity and would follow a quite simple equation, in which the ratio of object velocity v and light velocity c enters as square, subtracted from one, under a root. I will write down this relation, because I intend to then deduce important conclusions:

$$l = l_0 \sqrt{1 - v^2 / c^2} \qquad\qquad (12)$$

The quantity $1 / \sqrt{1 - v^2 / c^2}$ is called the Lorentz transformation factor, also used by Lorentz for his ether theory. The velocity of a moving object, for example a muon, would be v, and c is the light velocity.

When, before Einstein, Lorentz and Poincaré tried to fit the properties of an assumed ether to provide an ever constant light velocity, they first applied formula (12) which includes the Lorentz transformation factor. The formula, however, described the distortions generated by the ether, since the velocity v is assumed relative to the imagined ether (relativity, 2014). Einstein argued that the ether was not necessary, but now an irrational space-time concept provided for the required mechanism. When the general theory of relativity made the space-time even more complex and intricate, Einstein later himself invoked again the concept of an ether, which accounts for the manipulation of space and time through gravity (Einstein, 1920).

The mentioned experiment monitored the muons at a velocity of 99.8 % of light velocity and the observed contraction of their flight path was apparently by a factor of approximately 15. But let us ask a few questions.

Any object with a given length l on a reference frame moving at high speed v would shrink accordingly depending on its velocity and always in the direction of the relative movement. The question is, however, whether a rocket which is seen to be shrinking, does in fact shrink. In reality the strain experienced during shrinking would dismantle the rocket. Why should this happen just because someone from outside is looking at the object? And why is the shrinking process not seen perpendicular to the relative movement? Can an object simultaneously have different sizes? The answer seems to be that the shrinking is not real, but an illusion created by the finite light velocity which is used for measuring the size of the rocket. Then we do not see the reality, but an illusion, a kind of mirage, created by the finite speed of light which is used for measurement. This should also be the case for the discussed observation of the muon.

The theory of relativity was once designed to allow the same basic laws of physics to function in reference frames with different velocities. Of course, one would expect that these basic laws should not depend on the speed of signal transmission between the reference frames. One important property of the theory should be to always generate the same light velocity. Let us now simply perform a small intellectual experiment and assume that the object would travel at a velocity v close to light velocity c. Formula (12) shows that in this case the length of the object would go towards zero, because the ratio v^2/c^2 would go towards 1. The object would simply tend to become smaller and finally disappear. Now the light velocity c for signal transmission should be modified. First we will increase the light velocity towards infinity. Such spontaneous information transfer via light signals at extremely high velocity would seem to be ideal, since information would be momentarily present for a distant observer. This is not entirely unrealistic since quantum theoreticians are even discussing infinitely fast transfer processes within quantum systems.

With an infinitely fast transfer of information (c --> ∞), the root in formula (12) would go against 1 since v^2/c^2 would go against zero. As a consequence, the length l of the object at high speed would remain unchanged. The length l would retain its original length l_0. This appears to be an astonishing result since one could consider information transfer at infinite velocity as ideal and the most reliable information transfer. However, no relativistic change in length would occur.

The effect seen is the Lorentz contraction which is the phenomenon that a moving observer measures a shorter distance between two points than a resting one. From both points he needs a light signal. After he monitors the first one, he is already approaching the second when the corresponding second light signal arrives. The ratio between the velocity of the observer and the velocity of light affects the measured distance.

With an infinitely fast signal transmission, there is, of course, no time spent for relative movement. No length shortening is observed. And the now varied velocity of light c enters the theory of relativity indeed as a signal velocity determining the so-called "time axis" ct (light velocity c multiplied by time t, which results in the dimension of length) of the four-dimensional space. The remarkable result, when assuming an infinitely fast signal transmission, is that the length of the object, as determined by formula (12), does not change even at a very high relative velocity v, which could approach the present light velocity c of the measured object. The theory of relativity would simply not show the well-known dilation effects of length (and time). A rocket would not be dismantled by shrinking. This indicates that the relativity theory, in fact, shows a manipulated reality which is caused by assuming signal transmission at the finite speed of light, which is "c". Such a conclusion is also supported by the fact that a rocket that has suffered a shrinking process, due to a given velocity v, would regain its original length when landing.

Let us now change the velocity of light c in formula (12) against zero. In this case the value under the root first becomes smaller than 1 and then negative with the length becoming imaginary. This obviously means that first l, the length of the object, becomes smaller than l_0, the original object, and then a measurement of l_0 becomes impossible. The velocity of the measurement signal is then too slow.

What can we learn from this thought experiment? We can learn that the velocity of signal transmission c changes the length of the object moving at velocity v. Why is this so? If we measure the length of the object from three spaceships which are all at different velocities and positions, we would measure three different lengths (fig. 26). How can one and the same object simultaneously have three different lengths? Obviously we are not measuring the real object, but the object as it is seen manipulated and changed by a peculiar property of the information transmitting light signals. One is measuring a kind of phantom or a mirage. Physics relies on measured information, which is all right, but the measurement should not distort reality if we want to learn about this reality. And the reality should be the length of the object studied and the time experienced on it. And these parameters change depending on light velocity used for signal transmission. This is not what one expects to find. I have never encountered a discussion nor explanation of these unexpected properties of the well-known formula of relativity theory.

What is, in fact, reality? In the theory of relativity, time is an important factor describing relativistic reality. I believe, however, that clock time as Einstein defines and uses it is a factor which simply does not describe the time-dependent reality of energy converting processes in nature. As discussed before, time has to be transformed together with energy as action (chapter 11). Time alone in a four-dimensional space does not link causality for energy converting systems. It is a theoretical factor used for monitoring proceeding changes. It is without substance and one cannot use it to transform changes into other reference systems. One has to transform action, energy multiplied by time. What sense does it then make to manipulate time alone for a relativistic understanding of energy converting system?

Let us now look at another important formula, that of the total energy of a system. The famous relation $E = mc^2$ derived by Einstein from the relativity theory describes the energy of a mass m at zero velocity. It may be interesting to say that it has been shown that the derivation of this formula was not really based on logic reasoning but already presupposed in the derivation of the result (Ives, 1952). Classical derivations of analogue formulas have also been cited (Gut, 1981). Einstein speculated that the energy formula he derived for light (and which included arbitrary choices) may also be valid for any energy in general. Two years before Einstein an Italian geologist, Olinto de Pretto, published the same relation between energy and mass deriving it from other non-relativistic considerations (Olinto de Pretto, 1903). He saw the mass of uranium and thorium converted into energy during the radioactive decay. Before him, in 1900, Henri Poincaré apparently brooded over the same formula. I mention this not because I want to reduce the accomplishment by Einstein`s genius. The point is that we want to understand what this formula actually means within the relativity theory. In fact the formula $E = mc^2$ has nothing to do with relativity. It would be still valid if the four-dimensional space did not exist. Einstein had neglected relativistic considerations prior to obtaining the famous energy-mass formula. It can be derived with classic arguments only. The energy-mass relation cannot be used in support of the space-time concept, which is an important argument here.

If the resting mass m is accelerated to a velocity v, its total energy mc^2 changes according to the theory of relativity to:

$$E = \frac{mc^2}{\sqrt{1 - v^2/c^2}} \tag{13}$$

Let us again look at this formula while theoretically varying the velocity of light propagation c. When the velocity of an object approaches the light velocity c, the energy E goes against infinity. This has been verified with experiments in particle accelerators and is expressed by formula (13), since the value of the root approaches zero. We, however, immediately see that even at such a high velocity v of the object, we would only measure the energy at rest, $E = mc^2$ if the light velocity c for information transfer (hypothetically) became extremely large, approaching infinite velocity. In this case, the root would approach the value of one, but would the energy contained in the mass increase with the square of the new light velocity? For a widely changing light velocity c, a wide variety of values for the energy E could indeed be obtained. In fact, by changing the velocity for signal transmission c we could increase or decrease the energy content of a mass. This is in contradiction to our basic understanding of energy and of a signal transmitting information system. The reality of an energy system should not depend on the velocity of signal transmission from this system. In addition, observers on differently fast-moving relative frames would determine a different mass and energy of the same object (fig. 24). How can inertial and gravitational mass be equal in such an environment? In addition, the inertia of an object would be different in the direction of relative movement and perpendicular to it. A system should not have different masses, energies and inertia simultaneously and there is another obvious problem: energy cannot be created from nothing or be annihilated just by changing parameters for signal transmission and measurement. It would be a strange and bizarre universe which would follow such laws. The mathematics of the theory of relativity is apparently not consistent, which has also been argued by many critics (conservapedia, 2014; relativity criticism, 2014; Smarandache, 2013). It also means the above formula (13) does not describe the physical reality of an energy system, but it monitors the distorted information on such a system, obtainable via photons travelling at the velocity c. Are we really interested in such apparent distortions when talking about a moving space vehicle and about time and length changing on it? Or are we interested in what is really happening to the observed object?

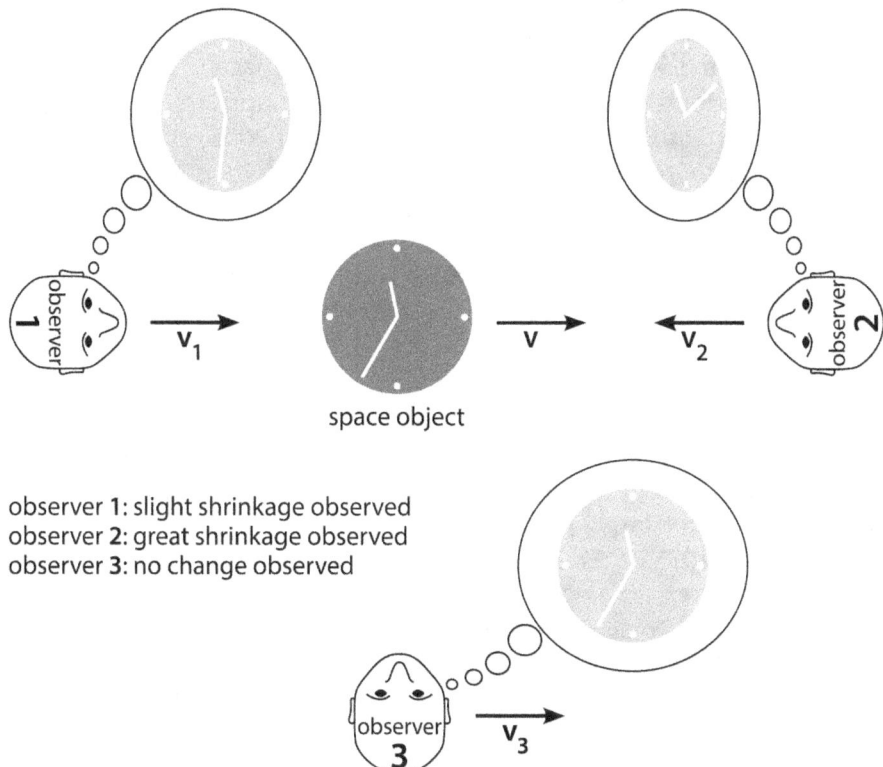

observer 1: slight shrinkage observed
observer 2: great shrinkage observed
observer 3: no change observed

Fig: 26. Three observers with different relative speeds will see a spaceship contracted to a different extent and see different timeflows. In the direction of movement the spaceship will contract and there will be a time shift, different for observers at different speed. Perpendicular to the movement, no length or time dilation is observed. What is the actual length and time of the space ship?

There is also a philosophical question to be raised. If the energy value is so drastically dependent on the velocity c of signal transmission, why then does the energy of the resting mass $E = mc^2$ just depend on this velocity c, with which the light is propagating now? This would mean, that by hypothetically changing the velocity of light for signal transmission all the energy contained in a mass and in the universe would also have to change. Where could this energy come from or where could it go?

To be precise: since the constancy of light velocity has been experimentally confirmed, and since the relativity theory considers exactly this fact, the theoretical results should somehow be mathematically useful. Also for the global positioning system (GPS) it is argued that it may need consideration of the relativity phenomenon. On the other hand, however, it is also denied that the GPS system has to consider relativity effects for practical operation (GPS, 2014).

We have to face the challenge that the presently generally accepted interpretation of the theory of relativity is entirely different from the conclusions presented here in this book. The hypothesis of a dynamic energy developed here does not need to make a statement on a property of space to justify a constant light velocity. The fourth dimension of space simply disappears. It is not needed. Light will transmit signals, and will arrive with constant light velocity without the need to assume a fourth dimension. Constant light velocity is a local property of the photon when measured. A four-dimensional space-time is therefore, to some extent, an irrational construction, a phantom, which does not exist in reality. And the formulas given by the relativity theory do not show the reality of the observed system, but the manipulation of reality using a finite speed c of signal transmission in a four-dimensional space. Even if objects moved close to the present light velocity, no relativistic effects would be seen if the velocity of signal transmission would be assumed to be infinitely fast.

But Einstein`s relativistic theories find a lot of support. Are they based on irrevocable facts? Einstein himself once commented to a journalist that it is the "mystery of non-understanding", which attracts many people who in fact do not understand (Pais, 1995). Some understood the physics and mathematics, but nevertheless could not accept the theory of relativity. An example is the famous French scientist Henri Poincaré. Until his death in 1912 he opposed this theory, as Einstein himself reported (Pais, 1995). Another famous scientist, Paul Ehrenfest, an Austrian professor teaching in Amsterdam, killed himself in 1933. In a letter to his colleagues, including Albert Einstein, he commented that he cannot go on teaching a science, which he cannot follow any more. Albert A. Michelson, who first experimentally determined the astonishing constancy of light velocity long before Einstein considered it in his theory, was not at all amused by the relativity theory. He said that he would prefer to believe that his measurements were wrong rather than believing this theory (cited in: Relativity theory, 2014). Other famous scientists that were witnessing the rise of the relativity theory and disagreeing with it were, for example, Ernest Rutherford, Robert A. Millikan, Ernst Mach, Wilhelm C. Röntgen and Nicola Tesla. I admit that I myself was surprised to have ended up with quite a critical view of the theory of relativity. I had no intention of criticising it, but the confrontation with it simply turned out to be a logic consequence of my considerations on energy. I stumbled over the possibility of explaining the constant velocity of light rationally via the information self-image of matter. The fact that the theory of relativity is so generally accepted in modern science and even communicated in schools as a pioneering achievement of human intelligence kept me from questioning it before.

What consequence can one expect when photons, via the information self-image of matter, automatically generate a constant light velocity in any frame with different velocity (fig. 25)? The consequence would be dramatic. There would no longer be a need for a Lorentz transformation factor. The relativity theories as they exist now would not be necessary. Light itself, and not space-time would impose an ever constant light velocity. Irrationalities with space-time would be eliminated. But what about the widely reported experimental verification of relativity theory?

A few additional words should therefore be said on the general theory of relativity because of its claimed great importance in the understanding of the universe. Its field equations describe the effect which matter and energy exert on the curving of space-time. They contain ten independent quantities, and, via the so-called energy-momentum tensor, also the mass density and the energy density (via $E = mc^2$). The general theory of relativity is based on quite complicated mathematics, but the availability of powerful computers has transformed it into a real playing ground for new concepts about space.

The questions raised in combination with the special theory of relativity consequently also apply to the general theory of relativity. If four-dimensional space-time does not exist, there is no need for equations "which describe the effect of energy and matter on space-time". Since they contain the energy density which depends on c^2 and in addition c, the velocity of light, to the fourth power (c^4), it should be asked anew how the velocity of signal transmission c can control energy content and fundamental properties in space. Again, with a signal transmission velocity c allowed to change from c is zero to c is infinite, widely varying space conditions and energy contents could be expected. And why is no conservation law for energy included in the theory? Philosophically the theory is puzzling and the mathematical framework is not entirely credible. One has the suspicion, as I already mentioned, that not the reality, but a space is described which is distorted by measurement expectations based on the assumed constancy of light velocity. This space has become so complicated in its properties that it has to be considered endowed with a special ether able to deal with adjustable space-time properties. In fact, Einstein himself considered this years after publishing the relativity theories (Einstein, 1920). I see in such a complicated property of empty space an irrationality. The imaginative physicist John Wheeler found suitable words to describe what is going on in space-time: "Matter tells time how to curve and space tells matter how to move". In fact, empty space dilates away from masses and contracts close to them. Time, on the other hand, contracts away from masses, and dilates close to them. This gives an approximate impression of what theoreticians expect to happen with empty space, when light or matter interacts with fast moving objects and light velocity, gravitation and inertia has to be adjusted by bending space-time. And there is another problem: the physical elements entering the general theory of relativity are time invertible, and the theory does not allow conclusions on energy conservation. However, sophisticated energy relevant conclusions are, as we will see, drawn on extremely dynamic events, the Big Bang scenario, the inflation of empty space, or on black holes and the accelerating space expansion.

The classical concept of a space which is homogeneous and contains nothing other than a void, has for a long time served as a helpful "symbolic form" for a rational imagination that fills it with objects. But a space-time, which can bend itself and manipulates the progress of time, does the opposite. Such a "symbolic form" does not create understanding, but confusion and non-understanding. Do we actually want such complications as a strategy to communicate science? When gravity can manipulate space and time, what sophisticated inbuilt properties must such a space have to accomplish that? What is in fact clock-time when space can accelerate and slow down clock-time around matter. For me, it is mathematical fiction. Instead of explaining a natural phenomenon such as the constancy of light velocity with a reasonably simple logic "symbolic form", space is simply assumed to master the complex phenomenon without giving clues about its function. A strange "symbolic form" was introduced. This is a form which is mathematically adjusted to explain desired functions, but nobody really understands how it should work. The "symbolic form", the irrationality of space-time, has basically become the faith that a brilliant idea of Albert Einstein is correct, the idea being to make empty space itself, and not a special ether, responsible for an always constant light velocity, for gravity around a mass, for inertia and for a manipulated timeflow.

Fact is that essential conclusions about the origin and dynamics of space were derived by stressing the mathematical possibilities of the general theory of relativity. Fact is also that highly dynamic, irreversible phenomena have been explained on the basis of time-invertible laws, considered in the general relativity theory. This concerns the expansion of space as formulated by Alexander Friedmann, the inflation theory as proposed by Alan Guth and Andrei Linde, black hole theories as discussed by Roger Penrose and Steven Hawking as well as Big Bang scenarios. The latter three phenomena are dealing with so-called "singularities" of the general theory of relativity. These are extreme situations, which the theory permits under special constellations of conditions, where possible mathematical solutions explode into a huge peak or an extremely steep and deep depression. The large number of parameters in the theory allows the calculation of a huge range of possible scenarios. The singularities are even able to destroy space - time itself.

Measurement expectations are quite relevant for quantum physics, which already has a significant impact on space theories. It should be remembered that Einstein, during a discussion of the quantum interpretation of the Copenhagen school, annotated:" Do you really think the moon isn't there if you aren't looking at it? " (Pais, 1979). Is the experimental verification of the relativity theory as good as is widely claimed?

It should be made clear here that the general theory of relativity has passed several critical experimental tests and is still at the centre of very active research. Einstein himself proposed three experimental tests for his general theory of relativity: the "perihelion precession" of Mercury`s orbit; the deflection of light by the gravity of the sun; and the gravitational redshift of light.

Gravitational time dilation and a frequency shift were actually found, but other explanations than the general theory of relativity are possible and the accuracy of measurements does not for certain permit the exclusion of other theories (Ohanian et.al., 1994). In the case of the explanation of the anomalies in precession of Mercury`s orbit, it has been claimed that the formula used for explaining the anomaly had already eliminated any relation to the general theory of relativity through simplification. Indeed, the formula had been derived from classical considerations. It has also been claimed that the relativity theory cannot predict the anomalous perihelion precession of another planet, of Saturn (Iorio, 2009). However, research goes on. Recently, after a 45-year preparation period and a cost of 750 million dollars, the gravity probe B satellite mission delivered gyroscope precession data, the behaviour of highly rotating spinning tops in space. The precession of the axis of the gyroscope was measured and its deviation from a precise orientation was found to be 1.8 thousands of one degree per year. A full circle is made up of 360 degrees. The extremely small deviation was attributed to a gravitational "warping" of space-time. It was claimed to be in agreement with Einstein`s relativity theory (gravity probe B, 2014). Do such very small effects, fitted to a complex theory full of adjustable parameters, really demonstrate four-dimensional space-time? An example of the complexity of experimental verification of the relativity theory is the deviation of starlight by celestial objects. It was already known to occur from classical gravity considerations. However, relativity theory predicted a larger deviation, approximately twice the size. This was initially confirmed by the expedition of Arthur Eddington to a solar eclipse of 1919 and made Einstein and his theory famous. However, during a careful re-examination it was then discovered that the confirmation was experimentally doubtful. It actually took 50 years of debates and studies until apparently a final agreement was reached in favour of the relativity theory. But is it the only correct explanation and what role does the adjustment of parameters that enter the theory play? Is such an extremely complex theory, supported by difficult to verify predictions and very small experimental effects, a theory which also produces many irrational and unreal results, an appropriate basis for explaining the origin, properties and fate of our universe? Nowadays, in science reports on television, when, for example, the doubling of a galaxy via a mirage effect of a gravitation lens, or the energy-mass formula is explained, Einstein's relativity theory is usually credited as the basis for this. The truth appears to be more complex.

Within the frame of the concept of a dynamic energy presented here, the effect of gravitation on light and matter will have to be evaluated in greater detail. Particles exposed to particle-wave duality in the form of relation (11) including energy in the quality of information will have to be examined when influenced by gravitation. It will be seen later (chapter 21) that gravitation can be identified with the information self-image of matter. One is ultimately dealing with an interaction of information with information. It is known that information is additive or subtractive. Effects can be enhanced or reduced depending on the information contained. There will definitely be an effect of gravitation on photon dynamics. A detailed theory will, however, be required for quantitative predictions.

Summarizing some of our conclusions: it should be pointed out that within the frame of a dynamic energy concept no four-dimensional space is needed to explain an always constant light velocity. In addition, the time axis of the four-dimensional space is a mathematical construction which, in my opinion, does not reflect physical reality of energy converting systems. Time itself has no substance and, as a physical factor described as $t = $ time, should not exist in nature. Time cannot be measured directly like wind or streaming water. It appears only in the form of action as a product of time t with energy E. Only for phenomena during which energy is not converted and time can be assumed to be independent of energy, the relativity formalism seems to play a certain role as a mathematical model describing the effect of constant light velocity on the measurement process. In fact, those phenomena which were recognized as being subject to relativity, elementary particles and atomic clocks, appear to belong to these systems. Seen that way, the Lorentz transformation factor describes a distortion of reality via a measurement which uses light or electromagnetic waves in general. When a moving observer sees a shortened length, this does not mean that the length of the object really is shorter. To me, the special und general theory of relativity appear to be just mathematical models of a physical situation of limited practical relevance. They are in part inconsistent and misleading, because they communicate a reality which depends on the speed of signal transmission. This would seem to be acceptable for quantum physicists who, basing on the Copenhagen school, believe that measurement only shapes reality. But if the measurement procedure, via a light signal with limiting velocity, is creating understandable distortions and irrationalities, this should disturb a naturalist, who is interested in reality. The relativity theories, in addition, do not respect energy conservation, and they describe nature via an ordering parameter, which is artificial and not related to reality, a time or clock time, which is independent of energy. How can such a parameter be expected to correctly describe and manipulate reality? And there are too many adjustable parameters involved, too many solutions of the relativity equations which have no practical relevance. There are also breath-taking singularities where extremely small parameter changes cause huge variations of results.

These considerations show that the well-established theory of relativity, which kids nowadays sometimes have already to face in secondary schools, can be questioned. And it can already be questioned on the basis of simple reasoning: because a most astonishing fact is that the theory of relativity, which is based on an entirely time invertible physics, is presently used to describe and justify extremely dynamic phenomena such as the Big Bang explosion, the consequent inflation of space, Black Holes, or the effect of accelerating space expansion. Does this really make sense? It should not. What can reversible physical laws tell a scientist which explains imagined highly irreversible phenomena?

Given the present general acceptance of the relativity theories in science and the publicity present in the media, this is quite an astonishing result. It is a result which was not at all expected at the beginning of my study. It resulted from the need to consider energy in the form of information for the explanation of a dynamic particle wave-duality (equation (11)). Such critical conclusions make it easier to attempt a replacement of the space-time theory of constant light velocity with an information self-image explanation of constant light velocity. Many inconsistencies, irrationalities and contradictions would simply disappear.

Some readers may feel that challenging theories by Einstein is arrogant. I disagree. What is being presented here is just an intellectual experiment and Einstein loved to engage in intellectual experiments. Anyway, until his death, he was not sure about the correctness of some of his conclusions. His formulas claim a non-imaginable four-dimensional space, irrational time travel, and an empty space, which can manipulate the size of objects and the behaviour of light and time near matter. This is the reason why they had to be analysed and questioned here.

19 Time travelling: endless paradoxes

There are additional mysteries derivable from the theory of relativity. When Δt is the time interval between two local events at the place of observer 1, the theory of relativity predicts time dilation of

$$\Delta t' = \frac{\Delta t}{\sqrt{1 - v^2 / c^2}} \qquad (14)$$

for an object moving (observer 2) with velocity v relative to observer 1. The duration of the clock cycle is found to increase when the velocity v approaches light velocity c. The clock cycles become infinitely long and time stops for a particle travelling at light velocity, such as a photon. This means that the moving clock is running correspondingly slower and finally stopping. This time dilation formula was originally derived and applied by Voigt, Lorentz and Larmor between 1887 and 1897 long before Einstein's relativity theory. But they were assuming an ether for the propagation of light. Later Einstein showed that the formula could be derived from the relativity concept and the assumed constant light velocity alone. It seemed to open the possibility of objects and organisms travelling in time.

Few science - related phenomena have stimulated human phantasy more than time travel. An incredible amount of scientific articles, speculations, books and movies have been dedicated to this subject. In any case. money can be earned with mysterious phenomena. Today most people seem to believe that somehow it may work. Interest started with the publication of Einstein's relativity theory in 1905. It predicted that a clock travelling fast from a reference point to a far destination would go slower according to above relation (14). Since this would also be the case for the return trip of the travelling clock the time would pass more slowly. In 1911 Einstein extended his conclusions to living organisms. He speculated that at their return, the local organisms that did not travel could already have been replaced by subsequent generations (Einstein, 1911). This interpretation survived and grew further in many speculations. The identified phenomenon became a more profound consideration later in 1949 for the creative mathematician Kurt Gödel. He, a friend of Einstein, dedicated it to his 70th birthday (Gödel et.al., 2003) (Yourgrau, P., 2005). In his world model, the "Gödel-Universe", he found a new solution to Einstein's equations describing the general theory of relativity. However it violates Mach`s principle. It rotates and shows time-like closed curves, world-lines, suitable for time travel. Time travel is thus compatible with the theory of relativity. Violations of causality are evidently possible in this model. However, the most perturbing result of Gödel's study was that, when time travel is possible, time itself, as we experience it, cannot be distinguished any more. The relativity model, aimed at better understanding a phenomenon like time, also leads to an answer which denies a meaning to time.

This supports our above statement that the time parameter, t, is just a mathematical construction and alone does not reflect a physical reality of change. It can monitor change, but does not implement change. This, however, would be necessary if realistic situations are to be examined and described. The time-experiencing reality in nature is action, energy multiplied by time.

Einstein said that time is what one reads from the clock, but he also said time is an illusion since one would not know what time to choose due to the phenomenon of relativity. Also, travelling to the past would produce confusion. Is one travelling to the past which really existed, or travelling to another past where the time traveller arrives and where he is included into past reality?

Again, also in formula (14) which describes time dilation, a hypothetical increase of light velocity towards infinity would eliminate this effect of time dilation, even for very high velocities v of the objects. This is puzzling and strange. In a universe of simultaneity the source of irrationality would simply disappear. It describes the situation of an absolute time in space. This indicates that with the relativistic puzzles we are indeed dealing with a problem of measurement, an effect of the finite light velocity which interferes with a nearly comparably high travelling velocity and manipulates reality. If the timeflow of an object is measured from three different spacecraft with different velocities and positions, three different measurements of time dilation will be obtained (fig. 26). Can one and the same object be subject to three different time flow rates? Then one is not measuring the real time of the moving object, but the time modified by relative velocities.

I also see an additional problem for time travel, which I have never seen considered. When a person is time travelling, he will activate a lot of new information. He will be recognized and observed by other people and he will interfere with their regular life. Information has an energy content and cannot be activated without a turnover of energy. And there may be other energy-consuming activities a time traveller may engage in. The arrival of a person in the past or in the future will definitely need additional energy. Where does it come from in a world which has long ago completed energy conversion activities or in a world which does not yet dispose of energy for changes?

Within the dynamic energy concept proposed here time cannot travel alone. Energy transfer would have to accompany time travel, but energy multiplied by time gives action. And action does not change when transferred from one system with a given velocity to another system with a different velocity. The time experienced remains the same. In every relative frame life would proceed in the same way and at the same rate. How then is time travel possible?

Reflecting on time travel has indicated many kinds of paradoxes. A well-known paradox is the grandfather paradox. A time traveller travels to the past and prevents his grandfather from fathering children. The time traveller could consequently not have been born, but he exists. A number of conflicting theories discuss the possible outcome of such an intervention. They range from the "multiple universe hypothesis" where the traveller ends up in another universe, to the "choice timeline" where the traveller changes the course of history, to the "destruction resolution" where the time and space concerned is destroyed (time paradoxes, 2014). Another time travel paradox deals with information. A time traveller travels into the future and reads of an important scientific discovery in the past. He learns about the scientist, who made the discovery and travels back to meet him at a time before his discovery. Now he gives him all information for the pending discovery. The paradox here is that neither the time traveller nor the scientist made the discovery, but the information was generated. Was it generated from nothing? This is not possible because information involves energy and energy cannot be created from nothing, at least not in the rationally understandable world discussed here.

Should we believe in the relativity theory or should we knock down a pillar of present energy understanding?

The dynamic energy hypothesis gives an entirely clear answer.

If a person wants to travel in time t he is faced with the problem that his organism converts energy E(t) and goes on in time. Time for him cannot travel alone, but only in alliance with energy in the form of the product E(t) times t, E(t)t, because both depend on each other. This product is action and is, even according to the theory of relativity, not changed when transferred to a moving reference system. A person on a fast moving object would live and progress in age exactly as we do in our world. Time travel for energy consuming objects does not exist.

If, on the other hand, a quantum system, for example an elementary particle, is involved, the situation changes. In a quantum state, energy is not converted and no time advance created. Here, mathematically, energy and time are not related and could be transformed independently within the theory of relativity. Time dilation here is actually possible and this is apparently what can be observed with muons, with travelling atomic clocks or particles in accelerators. However, time dilation is only present as long as the quantum system exists. This may not reflect an ultimate reality, but a measurement paradox, generated by the finite speed of light, which limits and manipulates the transfer of information. But it should be remembered that for a quantum physicist only measurement creates reality. He defines reality that way. My argument is underlined by the observation mentioned that in the time dilation formula (14) itself, time dilation would disappear if we used a more ideal hypothetical mechanism of information transfer than light: an information transfer at nearly infinite velocity, a speed of light c which approaches infinity. In this case, the root in formula (14) would become 1. We would not see any time dilation.

The conclusion that we are dealing with a measurement paradox is also convincing for the following reason. If today a single high speed particle among the uncountable ones from space or from particle accelerators travelled into the distant past to induce a relevant mutation in our past genetic code, evolution could be changed. How then could the present state of evolution be explained on the basis of past evolution? In fact, in order to change the path of evolution, the particle would have to produce an impact, it would have to generate energy conversion. In this case, action is generated and with action occurring, time travel is not possible according to my reasoning.

20 Elementary particles and fundamental forces

The hypothesis of a dynamic energy also allows quite precise new conclusions to be drawn about the existence of elementary particles in physics. An energy, which attempts to decrease its presence per state will actually engage in such reactions and drive time dependent processes. What will happen is that action (energy multiplied by time) will occur in the form of individual elements or cascades of elements of action. Since energy is assumed to be an oriented quantity, and also time becomes oriented, action is equally an oriented quantity.

This is not the case in present physics even though dynamic processes such as a stone rolling down from a hill (fig. 16), are explained using a dynamic principle, the principle of least action. The stone will follow the path of minimum action. This law has a general significance in all fields of physics. The difference in my approach is that energy, time and action are not mere numbers (so called "scalar" quantities), but they are oriented quantities (vectors). This has the consequence of there being an oriented time in the flow of elements of action. It has been shown before (chapter (9)) that a dynamic, time-oriented energy can be deduced from infinitesimally small integrals of action, which also have to become minimal. Evidence could even be given that the dynamic energy postulate and the principle of least action are equivalent. Why this has not been recognized before arises from the fact that energy was defined to be a mere number and fundamental processes to be time-invertible.

The parameter time is formally deduced by dividing action, energy multiplied by time, by an energy. If there is time, there is a "before" and an "after". In this case feedback processes are possible. And feedback processes are able to generate self-organization mechanisms. This will happen in particular when large quantities of energy are involved. Self-organized mechanisms are deterministic processes generating diverse well-known phenomena. Among them, chaos as well as structure and order generating processes are the most prominent. They are generated at the expense of disorder (entropy) in the environment. I am convinced that the large number of elementary particles and intermediate products are mostly products of self-organization of energy. During the degradation of energy into smaller units feedback processes will interfere and temporarily deviate this process in alternative directions. Self-organized order may be generated, stabilized matter produced. The mechanisms activated via the self-organization process will interfere locally with the energy degradation process. Energy becomes temporarily stabilized as matter. A similar phenomenon will occur as when water is poured over a hotplate. Only part of it will immediately vaporize, but water droplets will form and violently move around on the hotplate. We are talking about the well-known Leidenfrost phenomenon. In a similar way elementary particles could form from energy, temporarily stabilized via self-organization. These particles can be smaller or larger, can have electric or magnetic properties, can have different lifetimes, all will be consequences of self-organization of energy.

Such a dynamic approach to the understanding, the generation and existence of elementary particles is quite attractive and straightforward. It is a natural consequence following a fundamentally dynamic energy law (statement (7)). This law has already allowed a better understanding of the particle - wave dualism. Very small parameter changes can convert one structure into another. This is known from numerous examples of self-organization. The first structure could be the particle, the second structure, responding to a very small parameter change, the wave.

Let us first have a more general look at self-organization and begin with the phenomenon of chaos. It is now an already quite well understood phenomenon explaining many features in nature, such as natural river courses, turbulence in water, weather patterns or the sporadic appearance of insect pests like the gypsy moth. Here, very small parameter changes can generate very large consequences, so that an outcome can often only be described statistically. The relation between chaotic and (ordinary) statistical events has, in fact, been reviewed in greater detail. Berliner (1992) comments: "It is important to first decide if such a question (whether a phenomenon or data set is deterministic yet chaotic, or random) makes sense. I have already asked the question whether or not "ordinary" randomness is simply the result of uncertainty in the presence of determinism". This underlines the understanding that statistical processes arising from deterministic chaos cannot be distinguished from statistical processes expected from classical randomness. What can be learned from that? When a nuclear particle decays into fragments and energy, with statistical probability this can still be a deterministic process. And it is definitely one within the dynamic energy hypothesis. It is simply a chaotic process, a phenomenon of self-organization leading to chaotic destruction.

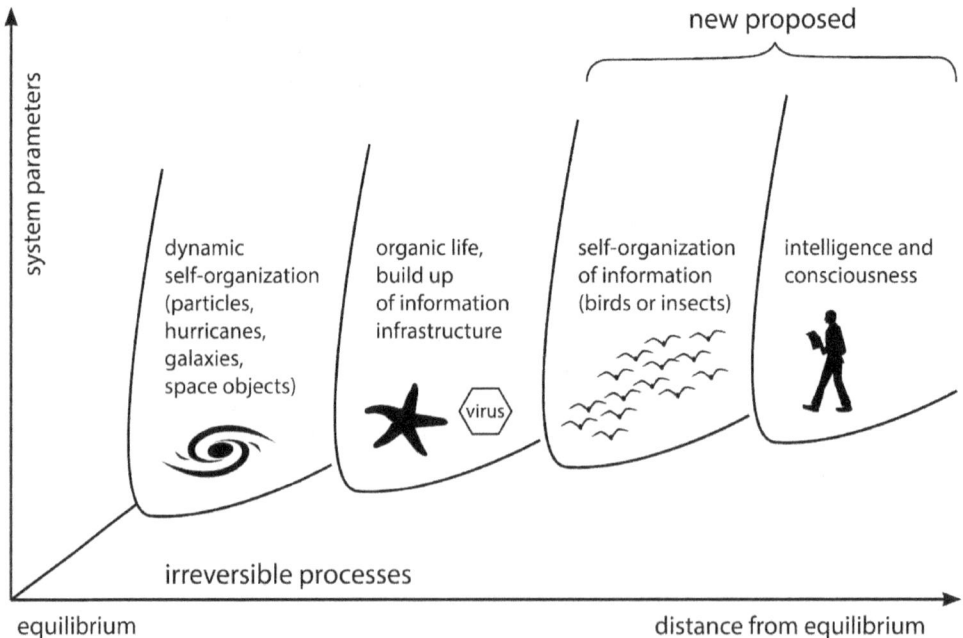

Fig. 27. In a non-equilibrium world the distance from equilibrium is important. It is close to equilibrium that irreversible processes start. But from a larger distance systems begin to self organize and support phenomena such as structures of galaxies or organic life. Then information structures will develop and may self-organize to yield a higher hierarchy of computation (see later).

Quantum theorists insist that such statistical processes of nuclear decay do not follow causal laws. There should be no cause for a quantum process. A particle is expected to decay without any reason. Non-causality is postulated to exist. Since classical quantum physics does not know time orientation, there is, of course, no room for deterministic chaos. Statistics from deterministic chaos is not part of quantum physics, but my approach of a dynamic energy allows it. It thus eliminates another paradox of quantum physics: non-causality. Einstein was right in claiming "God does not play dice". It is true that in this and other cases only statistical answers are possible. Here quantum theory is right, but the statistical response derives from a deterministic process. There is a cause. It is a process which may not yet be well understood, since a detailed theory still has to be developed, but it is deterministic.

Let us now look at other products of dynamic self-organization, which may occur during nuclear decay processes. It is a straightforward reasoning that the particles generated during a nuclear decay will also be the product of self-organization of energy. Particles and matter as self-organized energy? Since the particle processes studied in powerful accelerators usually occur far from equilibrium, one is actually dealing with processes far from equilibrium.

Science is today talking of a zoo of existing elementary particles. It appears that the number found up to today has reached 61. The recently found Higgs-particle, detected in collisions of protons accelerated to nearly light velocity, has stimulated much excitement. Its discovery merited a Nobel prize, even though the particle showed a lifetime of only 10^{-22} seconds (1 second divided by 1 with 22 zeros). This is an incredibly short time. How can such a transient particle be so important? It was expected to represent a field in space, which is attributed to generate the mass of particles. The Standard Model of energy and matter, which has grown and was modified in many steps (Standard model, 2014), is based on kinematics, which deals with motion of bodies considering their causes, and on interactions. These interactions may be electric, magnetic, gravitational or are activated via force-mediating particles. Mass is expected to essentially originate from the large bonding energy between quarks, the building stones of nuclear particles, but an invisible energy field, the Higgs field, apparently supports that. It is like a strange, modern ether in which particles are expected to move. It is an additional serious complication within the already so complex space-time web. Elementary particles are thought to experience a drag while moving in it. This is explained as the generated mass and the Higgs particle is assumed to be associated with the field, but the Standard Model contains 19 adjustable parameters and is not really elegant. A very wide range of imagined reality can thus be probed and adjusted. The model thus appears to explain many properties of matter, with the exception of gravity.

If, in contrast, the particles were just products of dynamic self-organization, kinds of islands formed in chaos, or vortices in turbulent water, or drops jumping on a hotplate, a theory for understanding them would have to be entirely different. It would have to contain reaction rates relating to feedback mechanisms, quantities defining the distance from equilibrium, parameters explaining the state and reactivity of energy and matter during the reaction. The properties of these particles, the more or less stable islands within chaos, their size, lifetime, or interaction ability, would result from such self-organization processes. Everything becomes possible because energy conversion generates timeflow which is fundamental. Feedback mechanisms can thus occur. The question why different particles have different masses, still a mystery within the Standard Model, would pose no problem. Simply expressed, the procedure would in principle be the same as attributing different masses to different kind of virions, inactive viruses.

In any case, the self-organization processes involved in the formation of elementary and transient particles would have one interesting and important consequence: matter is equivalent to energy since matter in the form of particles would be a product of self-organization of energy. A formula of the type of $E = mc^2$ is thus in principle confirmed. When, during a nuclear process, energy is turned over and decreases its presence per state, then new particles may form as stabilized energy, and waste energy will be released. However, that one type of particles formed, e.g. the short-lived Higgs or "God" particle, should determine the mechanism of mass synthesis, as assumed in quantum physics today, is not at all evident from the "dynamic" energy approach.

When elementary particles are assumed to be a self-organized product of energy, what does this actually mean? How complicated are such self-organized systems? In fact, a small parameter change could convert one self-organized structure into another, for example, a particle into a wave with all its different properties, or a stable nucleus into a decaying one. So far as I understand our present quantum physics cannot explain that.

Let us now compare elementary particles more in detail with tiny viruses, or better virions, when they exist outside living cells. Their size is between 20 and 300 nanometres (one billionth of a meter). They are not living organisms, because without the infrastructure of a host they cannot reproduce themselves and cannot convert energy, but they have certain properties and a program which they can put into action in a suitable environment. Elementary particles are similar to virions in their varied forms. They have specific properties, and when supplied with energy, such as in a particle accelerator, they can transform themselves. Tiny virions with their beautiful quasi-crystalline structure give an impression of the possible reality and minimum complexity of a self-organized elementary particle. When mass is formed from energy this way, one can definitively expect that the relation will be much more complicated than expressed via the proportionality factor "c^2" in the formula $E = mc^2$. Why also should an astronomical factor, the velocity of light, be the only determining factor for relating a self-organization process of energy to mass? However, it could be contained in the proportionality factor, because the constant light velocity is related to the information self-image of matter, which should be involved. It also seems to be far-fetched to imagine that such a short-lived transient mass-object as a Higgs particle may control self-organization processes of elementary particle formation. It is simply another product of self-organization far from equilibrium. There is, in addition, a philosophical question: Quantum physics is based on a time-invertible fundamental physics. Why should experiments, which are performed with extreme energy input far from equilibrium, provide relevant insight? Such experiments are much better fitting into a dynamic energy world in which a particle collider simply generates extended or new families of self-organized transient particles depending on the energy input and the kind of colliding species.

Now that we have started thinking about self-organized elementary and transient particles, we should also speculate about additional properties of matter: electric and magnetic properties, gravitation. Gravitation is, as we will see below, something special and different from electric and magnetic forces. If energy has the tendency to decrease its presence per state, self-organization mechanisms which support and improve that tendency are expected to evolve. Owing to feedback processes involved the energy turnover will be faster in self-organized mechanisms. They will consequently dominate and diversify into particles with additional properties. Gravity, electric and magnetic properties must be structures and functions of energy, which will improve the tendency of energy to decrease its presence per state. Indeed, the force of gravity aims at decreasing the energy per state by attracting matter. The same is true for electric and magnetic forces. Dynamic energy organizes itself through feedback processes and thereby develops structures and functions which support gravitational, electric and magnetic forces. These forces may thus additionally support the dynamic tendency of energy to decrease its presence per state.

The development of a theory of everything, a "Grand Unifying Theory", or GUT, is expected by scientists to come from an expanded Standard Model of particles in physics. This however is not at all obvious. First, it has to be said, that the GUT is not a complete theory, because it does not adequately consider gravity, and because essential parameters are deduced from experiments and not predicted.

An ultimate theory based on the concept of particles as self organized energy systems discussed here will have an entirely different structure compared to GUT, while giving reasonable explanations for size, life-time, gravitational, electric, and magnetic properties. Such a theory would, to some extent, be similar, but hopefully simpler than a theory describing different species of virions. Nobody knows at the moment, how complicated nature is on the level of self-organized energy and matter.

21 Gravitation: information self-image of matter

Since most of the mass consists of atoms which are in a quantum state and not in a state of decreasing their energy content, relation (11) should be applicable: energy in the form of particles tends to dilute itself in space, forming a wave, but information is activated to keep this process in balance. All of these together should have the ability to act towards a decrease of energy per state. Compared with a classical picture of matter, there is a difference. To understand matter rationally, it is not sufficient to describe matter, particle or wave, one also has to consider information on this matter. And this information should be real and somehow present around matter. What is it? This information self-image has an energy value and should therefore somehow be measurable around matter (fig. 19b). What energy-related phenomenon is found around matter? The answer is obvious. It must be gravitation. Can gravitational energy indeed be related to this information-mediated balance between concentrated and diluted energy?

There is an obvious analogy in our information system, based on television, cell phones and navigation systems. Here the information is contained in the electromagnetic waves which are transmitted to be present in the environment and to function via digital signals. As with gravitation, the distribution is not even. One realizes that sometimes when entering or leaving street-tunnels with a car and listening to the radio. The communication-mediating electromagnetic waves provide energy, and transmit information. The information, which was discussed as mediating activity between the particle and wave nature of matter must also be measurable via the energy which it involves. Gravitation, the mysterious force around matter, indeed involves energy, gravitational energy. It is therefore concluded that it is actually gravity which is related to the energy of information E_n in relation (11). The dynamic energy initiative has yielded a new interpretation of gravitation! It is not a force in the classical sense, but information with a well-determined aim, namely the aim of decreasing the energy per state. But it acts like a force. This deserves being explored more in detail, since gravitation is known to be something very special. And there is another remarkable fact: the information self-image of matter allowed us to understand an always constant light velocity and thus to avoid a counterintuitive four dimensional space. Now it offers an interpretation of gravitation, alternative to that generated by a four dimensional space.

When the information self-image of matter aims at decreasing the presence of energy per state, what happens when one attempts the opposite, to increase the energy per state, for example via acceleration of the same mass? One would expect a counterforce, a force that attempts to prevent an increase of energy per state. It can only be inertia. Gravity and inertial forces are determined to be proportional to the same mass or energy. They are forces active towards a decrease of energy per state or are activated in the case of violation of such a process. The equivalence principle, the observation that inertial and gravitational mass are identical, appears to follow in a straightforward way. The information self-image of matter, interacting with mass and energy, is doing the job. It is gravitation and acts on the particle while the particle decreases its energy per state. It is equally gravitation, which should matter, when the particle's energy per state is forced to increase during acceleration and thereby generates inertia, a counterforce. This conclusion results from the identification of the information self-image of matter with gravitation. The gravitation on the location concerned should, in fact, be the collective gravitation from the mass of the universe, which is responsible for inertia. This supports quite a vague idea by Ernst Mach on the origin of inertia, which was later discussed by Einstein, but only partially applied in the general relativity theory.

In the relativity theory the phenomenon of gravitation is explained differently and is much more complicated. The mass of a body generates a curvature of space-time. When moving along such a curved space-time, gravity and inertia are expected to become identical for a mass. This experimentally verified equivalence principle is a basic claim, which had been introduced into the general relativity theory. Space was adapted to act in such a way that the equivalence principle is fulfilled.

What is, in fact, gravitation and what speaks for its identification with the information self-image which mediates the particle -wave dualism of matter in the model presented here? In science today gravitation is still a poorly understood phenomenon and is discussed in a conflicting way, both physically and philosophically (Kohaut,Weiss, 2007). The "Grand Unifying Theory", GUT, for elementary particles cannot adequately assign it, because it seems to be something special. It will therefore be all the more interesting to find out what the dynamic energy theory has to say about gravitation as information self-image of matter.

Gravitation works for large masses of matter, but equally for elementary particles. But the difference in magnitude is large. Between two neutrons and between two weights of one kilogram the ratio of attractive gravitational forces is of the order of one to one with 54 zeros. Gravitation is nevertheless present everywhere. Gravity indeed supports the tendency to decrease the energy per state. It attracts matter and thereby decreases its potential energy and generates non-useful, chaotic energy, for example, heat. When a rocket accelerates at 9,81 meters per square second, an astronaut feels gravitation like on the earth's surface. He feels inertia, the resistance against an increase of energy per state. He feels it like a person on the earth's surface feels gravitation, the pressure to decrease energy's presence per state. When the rocket engine stops, the inertial force disappears. The energy per state does not increase any more. But if this pressure towards a reduction of energy per state exists why are atoms then not totally compressed?

Why are the negatively charged electrons and the positively charged cores of atoms maintaining a large empty space between them? There is, on the basis of previous considerations, an obvious reason why gravitation does not compress atoms and eliminates the enclosed empty space. The explanation for this not happening within the dynamic energy theory was already given before in chapter 12: energy has the tendency to decrease its presence per state and simultaneously attempts to dilute in space in the form of a wave. Information supports this process while also minimizing the information needed for inter-conversion of particle and wave. Quantization results from this attempt, atoms are forced into well-defined diffuse orbits of standing waves, which correspond to minimum energy-information conditions (fig. 18). Equation (11), by the way, would then describe the relation of gravity to quantum theory. It would describe a straightforward interrelation between the gravitation phenomena in space and the function of atoms and elementary particles. Such a relation has been intensively searched for in conventional physics but has never been found. The field of quantum gravity deals with that problem, aiming at a "theory of everything". String theory is an example of approaches, which have been followed. They have basically failed up to now, and from my point of view the reasons are obvious: quantum theory has to be modified and the gravitation concept of the general theory of relativity has to be abandoned. This appears to be a harsh statement, but is a necessary consequence of the considerations presented.

Within the model discussed here the unification of gravitation with quantum physics is a side product, however with an important new insight: the energy of information, E_n, identified with the energy contained in gravity causes quantization to occur (chapter 12), but is itself not quantized. Also the mysterious phenomenon of quantum correlation would be related to properties of gravitation and would suddenly become more transparent: when a quantum particle is split up to provide quantum correlated subsystems, the information mediating the particle wave dualism has also to be transformed and subdivided. This is not an easy job since natural laws have to be respected, such as, for example, the conservation of torque (spin). Some joint information structures are consequently maintained while the quantum correlated particles fly apart (fig. 28:). This makes quantum correlation intuitively entirely understandable. Since the particle - wave duality involves an information self-image, this self-image has to be dealt with when the particle is split up. Due to conservation laws only part of the original self-image can be split up and the rest will maintain joint information for the separating particles. Lacking a detailed theory it can presently, however, not be said up to which distance such information contact will be supported. However, for energetic reasons, I predict the reach of quantum correlation will not be infinite, as quantum physicists today claim, for quantum theoretical reasons (explained via relation (6)). All together the interpretation given here for quantum correlation would not contradict logics. The "spooky" action at a distance, in the words of Einstein, is a real phenomenon, and not really spooky any longer. Correlated particles are formed like married couples which maintain a reasonably strong contact via cellular phone even when temporarily separated. One just has to be prepared to accept that information on matter is part of quantum reality, and information also involves the possibility of communication between separating particles.

The fact that this very special phenomenon of quantum correlation can readily be explained is considered to be strong support for the discussed dynamic energy approach. Indeed, the very circumstance that quantum correlation exists also justifies the existence of an information self-image of matter for the original, undivided particle. The information self-image exists even before quantum correlation develops and provides the necessary information tool for implementing quantum correlation. However, one cannot generally say that matter has non-local properties. The phenomenon only occurs under special conditions, when particles and their information image split up and separate. But the fact remains that to rationally understand quantum phenomena both a description of reality, of particles and energy, and information on this reality is required (compare statement 11a).

This mysterious property of quantum correlation which may enable new technology of cryptography and secure telecommunication (Ma et. al, 2012) does not indicate an irrational foundation of nature, but follows from the simple basic and logic considerations promoted in this book. It is first a consequence of dynamic energy which imposes an alternating particle wave duality, and second a consequence of the information self-image of matter involved here. The quantum correlation phenomenon concerns the difficulty of subdividing it when particles split, but much more has to be learned about the nature of particle-wave mediating information or, as I claim, about gravitation. Quantum correlated particles are somehow engaged in an interplay with gravitation, which is the self-image of matter. Such a self-image in the form of information can obviously be read and handled by nature. In the case of quantum correlation over longer distances, man apparently has already started to interfere successfully, but monitoring and understanding information included in gravitation fields appears to be a long term challenge for science. Apparently, we have only just started.

Gravitation is still a relatively poorly understood phenomenon. In contrast to other forces in nature it penetrates all matter. A material body can also not be extracted from a gravitational field nor can it be shielded from it and gravity cannot be neutralized. It can, however, be partially or totally overcome by inertia. A high flying airplane in a sharp curve downwards leaves its passengers temporarily without gravitation. Within the model presented here this means that the drive to decrease the energy per state, and to give by to the attraction of gravitation, is just compensated by the drive resisting against an increase of energy per state, within the phenomenon of inertia. These two opposite drives are felt as forces, which in the described situation just compensate. In fact, however, these are two different communications of information, which are just compensating and neutralizing each other.

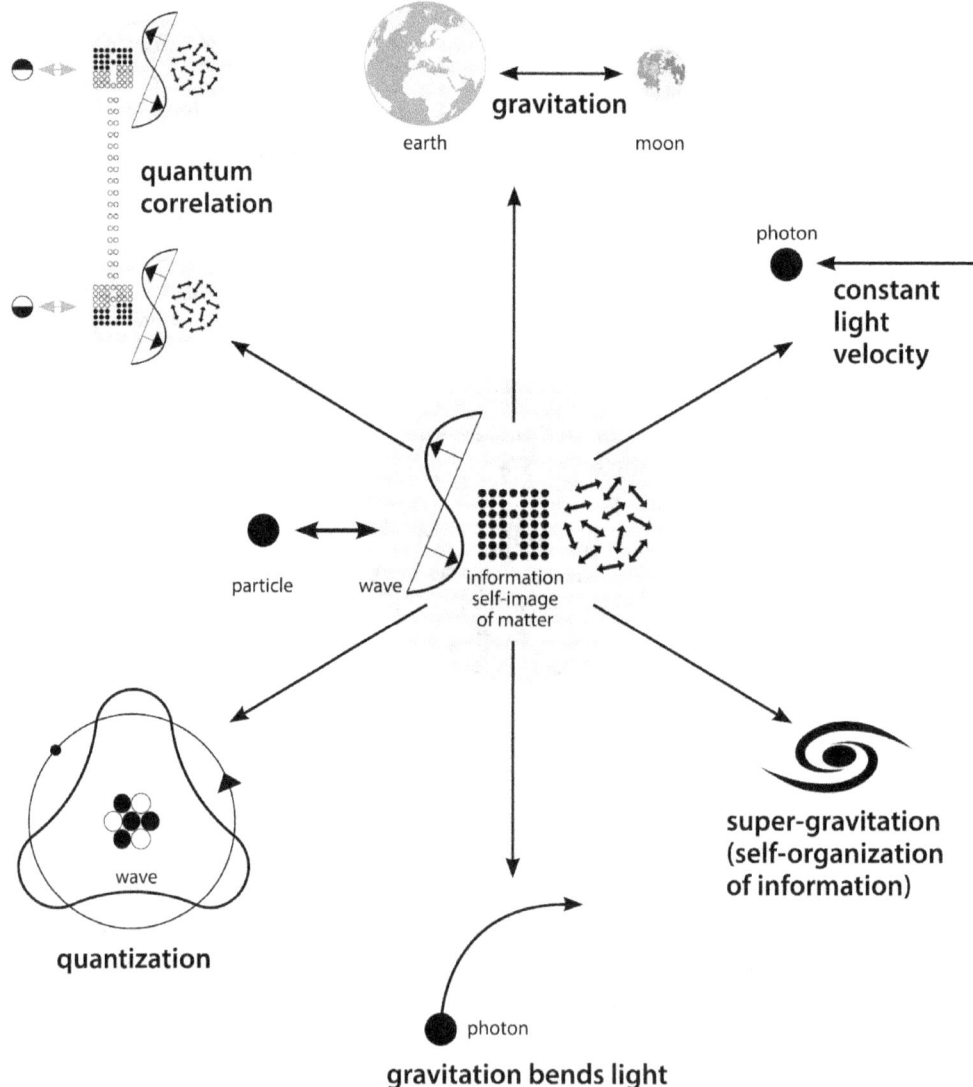

Fig. 28. The information self-image of matter, which mediates the particle-wave duality, was identified with gravitation, linking the quantum world with space properties. In addition, it is responsible for quantization, quantum correlation, and the force of gravitation acting on light and matter. Here it is claimed that gravitation can also self-organize to generate super-gravitation, a property actually attributed to dark matter.

Gravitation itself is caused by mass. Gravitation is, therefore, not the consequence of interaction of masses and it could until now not be explained on a more fundamental basis, which however the present effort claims. Gravitation needs mass, but it is not a consequence of action. Since action is the consequence of a cause, which requires time, gravitation should not be exposed to the flow of time. This is, in fact, also in agreement with the new model provided, which identifies gravitation with information. During the information-mediated particle-wave interchange no energy is converted and consequently no action is generated. Therefore, no time flows. In contrast, matter and energy themselves are exposed to changes and time. Gravity is quite a weak force compared to other natural forces, but it is far reaching and large in response to huge masses. But is it really a force? I claim it is information and the energetic aspect of information which aims at decreasing energy per state. In a gravitation field this is equivalent to an attracting force and explains why there is no repulsive counterpart of gravitation. It gives weight to objects, attracts them and causes them to fall towards the ground. From the time of Newton we know that when two particles with the mass m interact, the gravitational force between is proportional to the square of the mass divided by the square of their distance. The gravitational, potential energy is then inversely proportional to the distance between the two masses. The proportionality factor is in any case the gravitational constant. According to Newton it is a force at a distance, but nobody understands how such a far-reaching force may fundamentally work.

For Einstein, in his theory of relativity, gravity is not a typical force, but the consequence of the movement of objects through a curved space-time. It is an interaction at close distances. In fact, within such a picture, acceleration arises from a "curved" time. This is a time which changes its flow in such a way as to simulate acceleration. One should reflect here for a moment and try to imagine how a void with nothing in it can have such an elaborate property of manipulating time, which not even the most sophisticated technology can do. Nevertheless, the reason for planets cycling around the sun is consequently not attractive forces, but a curved space-time structure around the sun and the planets. The energetically permitted orbit for a planet is then its essentially elliptical planetary orbit. Photons, light particles, are also characterized by a mass and momentum and consequently deviated by gravitation. Einstein has simply transformed all the mystery about gravitation into a sophisticated four-dimensional space-time. All moving objects should be exposed to the same natural laws, including a constant light velocity. But Einstein`s theory of general relativity cannot explain the fundamental character and mechanism of gravitation. Empty space is just manipulated to impose the phenomenon of gravitation when masses are present. However, nobody can explain why and how empty space can implement such very special properties and this is a significant weakness. From the point of view of understanding nature, it is a clear setback. No detailed theory is imaginable that can explain such complicated behaviour of empty space. As mentioned, it is so complicated that Einstein himself, years after publication of the general relativity theory, was again speaking of an "ether" with special properties that should be able to do the job (Einstein, 1920). And

Einstein`s theory does not in addition enforce conservation of energy. It is also incompatible with quantum theory.

Why has the general theory of relativity then become so important in explaining gravitation and the dynamics of our universe? I have a personal opinion. First, nobody had an idea of how to explain an always constant light velocity differently. In addition, the mathematics of general relativity turned out to be so complex, that it was only gradually understood (Lichnerowicz, 1955). Many of its solutions turned out to be irrelevant, others difficult to interpret. Then new circumstances entered. Since the availability of modern computers, physicists can quite easily play with the complicated formula for general relativity and space-time involving ten parameters (General relativity, 2014). During my student time the information circulated that with five adjustable parameters an elephant can be depicted. Via these many parameters and the singularities which the general theory of relativity yields, such as extreme peaks and deep valleys, an incredible amount of possible scenarios can be described. They range from Big Bang details to space inflation and black holes. Only few solutions make sense, only a small community is able to interpret them and it imposes the identified "truth" on a wide public which cannot judge and criticise them, but sees the results in beautifully designed computer images on television. There is presently no evidence as to why the relatively inaccurate predictions of the general theory of relativity cannot be explained by other, less irrational theories, which at least would implement the conservation of energy.

The dynamic energy approach discussed here which has already helped to eliminate diverse paradoxes and irrationalities offers a much simpler alternative theory for gravitation: the energy related to information, E_n, involved in the interplay between particle and wave with the aim of decreasing energy per state. Information tells masses how to act. They act like systems guided by remote control. The mechanism is not action at a distance like Newton`s gravitation, and not action at a close distance like Einstein`s gravitation. It works as long as information is around and this means here gravitation. The technology works as if information is used to operate a system via remote control. An information signal must reach the moving object. Planets around the sun or satellites around the earth would thus follow the natural "remote control" mechanism operating via gravitation, forcing them to always fulfil the condition of minimum energy per state. They are exposed to gravitation and, accordingly, select their specific orbit. Wherever a planet or a satellite goes, the condition of a minimum of energy per state is fulfilled. It is ultimately the "dynamic nature" of energy, and its implementation, which is the secret of gravitation. Gravitation is neither a force, nor an interaction. It is information implementing a fundamental natural law which dictates a decrease of energy per state. Coming back to the example of a stone rolling down a hill, it can be said that gravitation forces the stone to always fulfil the condition of minimizing the presence of energy per state. We know now that we can also say that the stone follows the principle of least action in a dynamic way. The same happens with a satellite circulating around a star or a planet.

Let us use an additional check-list of properties of gravitation to verify in more detail whether these can reasonably support the claim raised that we are dealing with the information self-image of matter.

1) "Matter and gravity cannot be separated". This is evident from relation (11) since energy of information mediates between particle and wave form of matter

2) "Shielding matter from gravity is not possible. It penetrates all matter". There is no contradiction to be expected to this finding.

3) "Gravity effects can accumulate to large phenomena". There is no reason why energy of information should not accumulate in the presence of large quantities of matter. Information is additive or, if oriented into the opposite direction, subtractive.

4) "Gravity only attracts, it does not repel". This is evident since only attraction of matter can lead to a decrease of energy per state. Such a property makes gravity different from typical forces.

5) Gravity and inertial forces correspond to each other, but are opposite in direction. They reflect the action of masses from the universe. A mass feels all the gravitational forces of the universe. If a mass is accelerated, it acts against the gravitational forces of the universe. This is what the Mach principle claims and thereby rejects an absolute, empty space. In our picture, the following phenomenon would happen: when matter accelerates, the energy per state would increase, so that the energy of information, E_n, would counteract. This would be experienced as inertia. The effect would be supported by all masses from the universe felt via gravitation. Inertia would thus simply be a counteraction to gravity, which implements the claimed tendency of energy to decrease its presence per state.

6) If gravitation is not the consequence of a cause, then it is not exposed to a flow of time. Indeed, the modified particle-wave duality expressed in (11) does not involve a time flux, since energy only spreads and contracts in space, but is not turned over. It is an ongoing, eternal phenomenon and will be present as long as matter exists. As a consequence, there cannot be a spreading of gravitation when mass is localized, since spreading would involve time.

7) The general theory of relativity predicts gravitational waves. In spite of significant experimental efforts, they could not yet be confirmed. The identification of gravity with information mediating the particle wave duality allows a clear conclusion: information, and the energy related to it, is not expected to be able to generate waves itself. However, self-organized information, expected in a region of exceptionally high gravity, could, in principle, induce oscillating phenomena.

8) In a given gravitation field all bodies, whether as light as a feather or as heavy as a hammer, are exposed to the same acceleration. They approach the ground at

identical speed if air is not present to exert different friction. One could rationally assume that the information self-image imposes such behaviour, an identical acceleration. The information given in the same gravitation field towards a decrease of energy per state is simply identical for different objects. The force experienced is different, however, because acceleration has to be multiplied by the mass concerned.

The information self-image of matter could thus indeed be the origin of what is called gravity. It is not a force and not an interaction. It is information on the state and dynamics of matter, energy, which implements a fundamental dynamic law.

Interesting details could be added as additional support. Among paradoxes in quantum mechanics, the tunnelling phenomenon was mentioned above: the ability of particles to penetrate energetically too high an energy barrier. My rational explanation was that information partially penetrates the otherwise too high barrier and reassembles the particle behind. After identifying the information with gravitation, we know that this information can indeed penetrate matter freely- as gravitation. This can be valued as an additional argument for the consistency of the dynamic energy approach and its rational basis. The discussed quantum mechanical mechanisms, the differently interpreted particle wave dualism (fig. 19), the alternatively rational explanations for non-locality and quantum correlation (fig. 21), as well as of the tunnel phenomenon confirm the link between quantum mechanics and macroscopic gravitational phenomena. Such a link between gravitation and quantum phenomena could never be established in traditional physics. It is an encouraging success and should be the basis for exploring additional phenomena.

If we, for example, want to understand the deviation of light in a gravity field, or other gravity-induced phenomena, we would have to explore interaction possibilities of the particle-wave relation (11) with the perturbing systems. Tentatively I would say that, when gravitation (information) from a photon superposes gravitation (information) from a larger mass, a corresponding additive superposition should be expected. Light will, of course, react in response to external gravitation.

How could the explanation presented here for gravitation be experimentally distinguished from Einstein`s gravitation within the general theory of relativity? Information itself should not be able to develop waves. In this case we distinguish information (e.g. digital signals) from signal transmitting waves. The here presented concept can be falsified (Popper, 1978): it should be wrong if gravity waves could be measured. Such waves should involve also a periodical change of timeflow, which I consider utopic and unrealistic.

Later, when talking about the dynamics of the universe we will again stumble over the mystery of gravitation, the energy of information, which links concentrated dynamic matter and energy with distributed matter and energy of increased entropy. It will be learned that gravitation (information) could, under certain conditions, self organize into a phenomenon of super-gravitation. In this case, periodic phenomena would be possible, but they would not involve an oscillation of time itself, as claimed by the general relativity theory.

Since gravitation also plays a fundamental role in determining the dynamics of the universe, it may be concluded that the information involved in gravitation acts in a similar way there. It is information which should consequently control the fate of the universe. This will be considered later.

22 The cosmological redshift: a new interpretation

Quantum physics has had a tremendous impact on our present concepts of the universe. The main information we have from space arrives via light, photons. The properties of these photons and the general theory of relativity, which considers a space which implements constant light velocity, have significantly shaped the established "symbolic form" which has created the image of our universe: it is a four-dimensional space-time structure which is expanding dramatically fast. This has been deduced from the redshift of characteristic light from stars and supernovae in galaxies which is stronger the more distant the objects. The dynamics of space turned out to be quite complex. Today an accelerating expansion of our universe appears to be established. However, only part of the redshift is attributed to the Doppler effect (which is also known from the changing sound of a bypassing train). In this case the light source which flees at high speed deals with the light energy which is lost by the photon during the generation of the redshift. The energy balance is all right. The second currently assumed important contribution to the redshift comes from the so-called cosmological redshift. It is assumed to be caused by the expanding space of the inflating universe, of the void, itself. The photons are expected to experience the expansion of empty space after they have left the light-emitting stars. Their wavelength is assumed to be stretched by the void which itself is expanding, and they thereby undergo a redshift. This way they are expected to lose energy, but nobody really knows how empty space can do that and where this energy may go. Theoretically energy has to be conserved. But it is not. This is one of the irrationalities which at present shape an important concept of our universe. Another irrationality is to imagine that space itself, with nothing in it but light particles, can expand. Empirically, one is accustomed to increasing or decreasing a void space by adjusting the marking. However, stretching the void itself, which is expected to contain nothing, is not logical. It is just a mathematical manipulation, a speculation, which changes the energy of the photons involved without explaining how this is possible and where the surplus energy goes.

Experimentally the redshift from starlight is well studied and documented, because it is considered so important to understand the universe. All imaginable mechanisms were therefore investigated to explain the observed redshift. In a review Louis Marmet (2014) compared the incredible number of 59 different theories and concepts, which were proposed by scientists during years passed to explain the redshift of starlight. They consider a time-dependent distance, a time-dependent gravity, a time-dependent property of light, a time-dependent property of matter, a time- dependent geometry of space and time and many additional imaginable time-dependent phenomena as possible mechanisms for the redshift.

To evaluate the redshift, the relative difference of frequency between observed and emitted light wavelengths is described as a dimensionless quantity z. For the microwave background radiation of the universe, which in the meantime has also been well investigated, this z-value can be determined to be z =1089. This means that the shift in energy from the original photon to the microwave photon is 1089 times greater than the energy of the microwave photon itself. In other words, the microwave photons background carry only 1/ 1089 of the energy of the original radiation of starlight. Today it is understood that galaxies of the universe are expanding at an incredible speed. The highest redshifts for galaxies measured at a range of z = 6.96 to 8.6. The galaxy with the latter redshift is considered to be observed only 600 million years after the Big Bang, which occurred 13.8 billion years ago (Redshift I, 2014). Ten years earlier the record in z-values was z = 4.25 for galaxy 8C1435+635, located 13 billion light years away. Its escape velocity was measured to be 93% of light velocity (Redshift II, 2014). This is incredibly fast considering that on the basis of relativity theory, a huge amount of energy is needed to accelerate objects close to light velocity, which could only be reached with an infinite energy input. To reach such a velocity the energy contained in the mass of this galaxy would have had to increase by a Lorentz factor, which is approximately 2.7. How much is this? Let us use a comparison. Very little mass is needed to power a nuclear bomb. The Hiroshima bomb explosion from 1945, for example, consumed only 0.6 grams of uranium mass. This gives an impression of what a 2.7 fold mass increase of an entire galaxy with more than hundred billion stars means. And there are billions of galaxies seen and expected at correspondingly high redshift velocity. Where did this gigantic amount of energy come from? From a tiny Big Bang seed of the universe? Or did scientists calculate something from the redshift and the theory of relativity that is not realistic? There is another problem with the redshift of galaxies, which move away from us with velocities close to that of light. Let us consider two of those escaping into opposite direction. Then the relative velocity between them should exceed the velocity of light. This is, however, not allowed.

Can our dynamic energy approach bring more logic and more clarity? We have learned that subject to this dynamic energy concept, a photon tends to oscillate between the particle state and the wave state (fig. 19b). However, it has thereby to activate energy of information E_n to compensate entropic energy, E_e, which can no longer be used for work (relation (11)). This loss of working ability of energy, when energy density decreases, should interest us. The second law of thermodynamics and other facts from science have been mentioned as examples which demonstrate that energy loses its working ability when provided at increasingly lower density. Photons expanding from a star into space are no exception! They have a mass and the formula for the expansion of an ideal gas into a vacuum gives the amount of entropy increase (relation (4)) which is to be expected. Also the amount of entropic, non-available energy generated is well defined (relation (5)). In fact, in 1905 Einstein compared the formula for the entropy of a gas, expanding from a volume V_o to a volume V, with that previously derived by the German scientist and Nobel laureate Wilhelm Wien for expanding thermal radiation (fig. 18b) and found them to be identical. His conclusion was that electromagnetic radiation (e.g. light) does not have to be represented as waves, but can also be represented in the form of small energy particles (Einstein, 1905). They also appeared to be more appropriate for describing electron emission from solids. It is well known that this conclusion, derived from comparing published results, later earned Einstein the Nobel Prize. The light particles were later given the name "photons", but here a problem of irrationality arises. According to quantum mechanics, photons expanding into empty space cannot release energy in smaller quantities anymore. They have to somehow interact with other particles or with solids for energy release or with a gravitation field. When thinking in terms of photons, there is no way of getting rid of entropy, non-usable energy, during their expansion into free space.

Such an entropy liberation process would, however, be needed for fundamental thermodynamic reasons. Entropy production, which works in the wave picture of radiation, cannot simply be eliminated as a natural phenomenon because we look at radiation only in terms of particles on the basis of established quantum theory. Quantum physics, however, dictates such an impossible contradiction. This is a very important point in support of the line of thoughts developed here and again confirms the reason of inconsistency between classical physics and quantum physics: quantum physics does not properly consider the role of space for energy. It simply doesn´t care about useful and non-useful energy during the spreading and dilution of light quanta or particles into space.

Here the usefulness and correctness of my re-interpretation of the particle-wave duality (11) becomes evident. A particle constantly transforms into a wave and into entropic energy, but with the energy needed in the form of information for reconstruction into the particle form. With this involvement of an "information self-image" of matter, entropy formation during expansion of a radiation or quantum field can properly be considered, even when talking of photons. They can lose free energy during expansion into space, but this is not the only argument for entropy turnover during expansion of light into space.

It was already mentioned that spread radiation does not return voluntarily to an antenna, and that light at too low an intensity will not split water in a photo-electrolysis cell. When photons are emitted by a distant brilliant star, they start at high photon concentration. When arriving on earth they are highly diluted. Entropy turnover means the dilution and randomization of energy. If entropy is calculated for radiation leaving a star, one gets a logarithmic dependence on distance (derivable from relation (4)). Sound, earthquake strength, or the acidity of liquids are all measured in logarithmic quantities. Fractal geometries and self-similarity follow logarithmic rules. Logarithmic quantities describe the exponent by which a base has to be raised to yield x. They grow very slowly for large values of the variable (x). The energy loss of stellar light should approximately grow logarithmically if plotted against the distance of the light source. This energy loss of light during expansion into space is a more probable mechanism than an ever faster expansion of the universe. If, in contrast, one hypothetically assumed that photons at high density and very low density in space have the same working capacity, one would run into problems. There is definitively a difference. Information, and thus additional energy, would be needed to re-collect photons which are widely distributed from a star into space. All irreversible processes, by the way, lead to entropy production. And entropy production leads to a loss of working ability. Where does this loss of working ability in radiation show up in space?

Equation (11) suggests that on the basis of the new particle - wave duality used here, it first turns up in the two quantities E_n and E_e. Non-usable, entropic energy, E_e, shows up and the expanding photons may somehow get rid of part of it. It is important to note that the magnitude of entropic energy for expanding radiation will be proportional to the temperature of the environment. This is space, which can be as cold as -270.4°C or 2.72 °K, which is to be considered in formula (5). The energy quantities of non-usable energy from starlight should correspondingly be very small and are matched to the temperature of the universe. For individual photons RT has to be replaced by kT (with k = Boltzmann constant). The quantity kT is used as a scaling factor for energy which is relevant under the prevailing conditions. For cold space it can be calculated to be 0.240 milli electron volts. This energy value corresponds to radiation in the microwave region. This rejected waste energy has to show up again somewhere else in space. There must consequently be a great deal of energy somewhere in space which is detectable in the form of these very small energy packages. Where is this immense energy located in space which should have dramatically accumulated with time? There is only one possibility: the microwave background radiation of the universe. Even though this radiation has a very low energy content, it totals to such a huge amount that it represent most of the radiation of the universe.

The idea that photons lose energy during their long trip through space already goes back to Walter Ritz in 1908 (Ritz, 1908) and Fritz Zwicky in 1929. Considering the laws of quantum physics, the latter scientist assumed that collisions with small particles could decrease the energy of photons. In this case, however, the image of distant galaxies would be seen blurred. This, however, is not the case. The so- called "tired light" mechanism has therefore been discarded as an explanation of the redshift of stellar light. Our space telescopes see a non-blurred sky.

The idea presented here is different. A visible light particle has an energy of between 3 and 1.5 eV (electron Volt). A microwave quantum from the maximum of the cosmic background radiation has an energy of approximately 0.6 10^{-3} eV (0.6 milli electron volt). In fact, considering the measured redshift value of z = 1089 for the microwave background, one also deduces that the background microwave energy is at least 1089 times poorer in energy compared to light from galaxies and stars. The corresponding masses for microwave particles are too small to distract visible light particles markedly when microwave quanta would be released. In addition, a very elegant mechanism appears to be possible: the energy accounting for information, E_n, may simply deal with the need to set aside entropy by assembling microwave photons besides of photon radiation with a correspondingly smaller amount of energy. The entropy loss of travelling light expanding into space may thus be automatically accounted for and there may be no significant perturbation of the path of light.   And within the constraints of our fundamental dynamic energy statement, energy release would, of course, be permitted. The value estimated previously for released waste energy, for entropic energy, is actually of the same order of magnitude as the energy of the microwave background radiation in space. The empirical 2nd law of thermodynamics, prescribing an energy loss in the form of entropy, would thus be implemented. I am just advancing the hypothesis here that the light particles with their inbuilt tendency to decrease their presence of energy per state could get rid of entropic energy by appropriately tailoring the back conversion of energy in the form of information, E_n, into the particle form.

I again emphasize that on the basis of today's understanding of quantum physics this would not be possible. Photons cannot perform internal irreversible processes in the absence of interactions with matter or gravity. Photons also do not experience time in their own frame of reference, since they travel at light velocity. When time does not pass, energy cannot be turned over. Even if photons spread out in space, their entropy, disorder, would not change with distance. No irreversible process is expected during radiation transport and photons would still have their full energy when arriving somewhere. For this reason, scientists assumed a stretching of the void space which generated a stretching of the photon's wavelength, to explain the observed redshift. Here it is again seen how the inability of quantum physics to deal with the effect of space on energy became a decisive factor which gave rise to an irrational theory, that of the ability of the void - of empty space -to stretch.

The process of photon emission and photon distribution cannot be inverted without information and energy input. And when photons spread from a star into cosmic spaces, irreversibility increases. That this is indeed the case can also be visualized through a thought experiment in which the photons emitted from stars are used for energy gain. As shown in fig. 29, a solar sail could be applied to capture the momentum of photons in analogy to a piston exposed to gas.

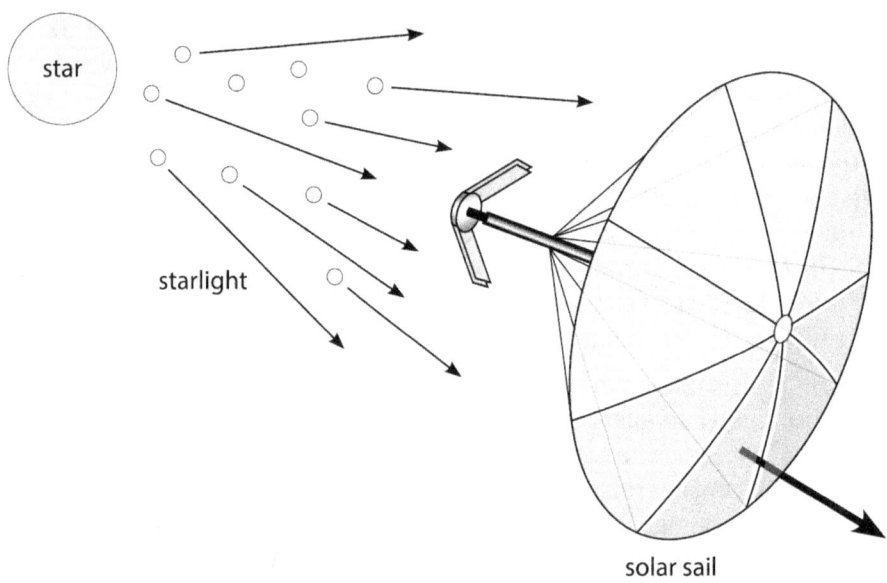

star

starlight

solar sail

Fig. 29. When starlight spreads into empty space, entropy is generated. A solar sail accelerated by starlight could harvest part of this spread- out energy. But if this energy is not harvested, it is converted into a non-useful, "entropic" form of this. Where is starlight entropy deposited in the universe?

Such sails represent an established propulsion technology that uses the radiation pressure of light to accelerate ultrathin mirrors to high speed (Solar sail, 2014). The solar sail could accompany the expansion of photons and gain energy. Useful work could thus be generated. The energy involved in driving the sail is not gained through an irradiation of light into free space. It then corresponds to the "non available" entropic energy abandoned during the photon expansion. In reversible thermodynamics, entropy generation during a sudden expansion of a gas into a vacuum occurs in the form of complicated processes involving heat turnover, acoustic processes and mechanical turbulence. How the entropic energy turnover of starlight actually functions will have to be investigated in detail. The most probable mechanism would, as mentioned, involve the back conversion of the wave energy into particle energy via information (relation (11)). Instead of one photon, two photons with very different energy values could be generated respecting the entropy law. For reasons of energy minimisation, the microwave photon would be emitted in the opposite direction of the star light photon. The low-energy photon would account for chaotic, entropic energy. The spreading out light waves would cover increasingly larger spaces, thus diminishing their energy content. Since the energy is inversely related to the wavelength, a redshift would be observed. The difference, entropic energy, would be liberated in the form of microwave energy, the background radiation.

In the theory of relativity, information transfer via light particles is assumed to be entirely reversible. I do not agree and I believe that our world is fundamentally irreversible and that also photons have to respect entropy laws.

Irreversible thermodynamics has definitively to be applied to starlight. The here proposed hypothesis of a dynamic energy underlines the irreversibility and also admits the concept of time. The photons arriving after a long journey from distant stars should not have conserved the working ability present at the beginning of their trip, but should have lost energy in the form of non-available, entropic energy. I simply claim that background radiation of space is not the echo of the Big Bang explosion of our universe, but the result of accumulated entropy production resulting from stars and galaxies. Possible, not all of the microwave background radiation originates from spreading starlight. The rest could arise from self-organization processes in space, such as galaxy formation and function. Such structures generate local order at the expense of entropy generation. This entropy should also contribute to the microwave background. It can be considered to be the entropy fingerprint of the universe. It is the deposit of "chaotic energy", of waste energy. The tiny deviations in temperature of the order of 30 parts per millionth of a degree are not the origin of future galaxies, as seen today, but the entropy mark or entropy footprint of galaxies that once existed and of those which still turn over energy and generate entropy (fig. 30).

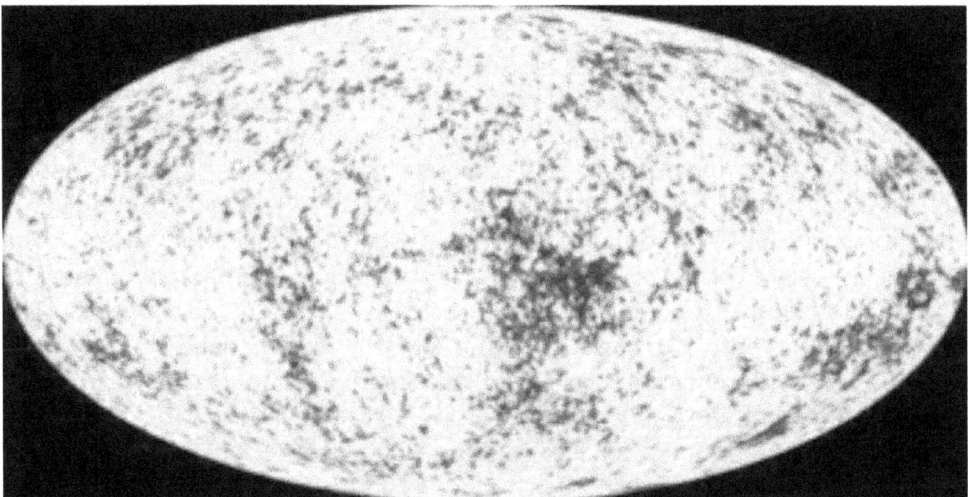

Fig. 30. The microwave background of the universe has been highlighted as the echo of the Big Bang explosion of an expanding universe (measurement by the Wilkinson Microwave Anisotropy Probe (WMAP), Credit: NASA). Even the observed deviations in temperature of the order of only 30 parts per millionth of a degree have been considered highly significant for galaxy formation. Here it is just seen as a gigantic fingerprint of entropy production, accumulated "waste" energy of spreading starlight and of galaxy activities.

The concept presented here makes a great difference to the present explanation of the microwave background, which is considered to be an echo of the Big Bang explosion. This presently widely accepted explanation is based on irrational assumptions and violates, as we have seen, known laws of physics. My interpretation claims fulfilment of a basic law: the increase of entropy during the spreading of particles (photons) into space and generation of entropy during the build-up of galactic order.

I admit that this is a somewhat courageous statement, because the current explanation for the microwave background is entirely accepted by the science community. But I insist that the alternative model presented here is much less speculative than the Big Bang explanation given for the microwave background radiation.

Today the microwave background radiation is declared a snapshot of the universe at an age of 380,000 years, calculated as the time (what time was actually running?) at which the universe became transparent for light. Now it is seen as black body radiation with a temperature of 2.72 degree Kelvin, corresponding to approximately -270.4 degree Celsius. It is assumed to be the consequence of space expansion. Light waves became stretched, because the void expanded until they ended as microwaves. Nobody knows where the energy difference remained. The distribution of microwave intensity shows some slight ripples, which were interpreted to have given rise to the presently existing web of galaxies and dark matter. These ripples were of the order of only 30 parts per million. They are of the order of micro-Kelvin and are absolutely tiny. Intensive research based on space telescopes is going on to shape these micro-riddles into concepts explaining evolution and the distribution of galaxies. There are astonishing assumptions behind this elaborate microwave background theory. Why is just this snapshot of light emission from the 380.000 year old universe responsible for the background radiation and what happened with all the light emitted later? To explain the even angular distribution of the radiation in space, not expected from an explosive Big Bang event, it had to be assumed that immediately after the Big Bang, the empty space itself expanded to a huge dimension. The phenomenon was called inflation. This is an incredibly arrogant scientific statement. We will later look at it more in detail. Afterwards, when the temperature of the young space arrived at 3000 degree Kelvin, the ordinary accelerating expansion of space continued and gradually "stretched" the wavelength of radiation to yield the presently observed pattern of microwave background radiation. The latter phenomenon is not compatible with the energy conservation law, since stretching the wavelength of a photon decreases its energy content. And there is no law which permits the stretching process and yields thermally equilibrated microwave quanta. As photons travel at light velocity without time, and since they cannot undergo irreversible processes in the absence of interaction with matter, standard quantum physics runs into serious problems. Where did the excess energy go and by what mechanism?

There is no obvious quantum mechanism imaginable. And nobody really seems to care.

This is also true for two well-established facts about the redshift, which do not at all fit into the present explanation of a fast expanding universe. One is that cosmic objects were identified, which had an obvious identical location, for example a quasar within a galaxy, but showed a strikingly different redshift (Arp, 1987). Another astonishing effect is a periodicity of the redshift seen when large numbers of bright galaxies are evaluated and their redshifts plotted depending on their determined escape velocity (in km/s). Periodicities of 71.5 km/s and 37.5 km/s have been observed in the environment of dense galaxy clusters and field galaxies respectively. A corresponding periodic clustering of galaxies was discussed as a possible ultimate cause, but periodicities of redshift are also seen within individual nearer galaxies. Such redshift behaviour cannot be caused by space-time expansion or simple light propagation through space (Paal,1970) (Tifft, 1973) (Napier et al., 1997).

My proposal that the microwave background derives from entropy generation from expanding starlight has to be confronted with the same question. How does starlight lose entropic energy? How can the energy loss be periodic? But there is the advantage that on the basis of the dynamic energy hypothesis proposed, one is now dealing with a world driven by energy and subject to irreversible thermodynamics. This has opened new opportunities for explanations. In addition, time can become active allowing time-dependent processes, which is not the case for the classical photon which travels without time and can lose energy only when interacting with matter.

It is remarkable that the answer to the question posed has already been given in detail by the proposed new dynamic energy theorem (6): "energy has the tendency to decrease its energy per state, within the restraints of the system". Applied to a travelling photon, what is observed in space is exactly what happens: photons lose energy, and the changing restraint is the entropy law. Their frequency is redshifted over long travelling distances. And the longer the photon travels, the more redshifted it is. Our new "symbolic form", applied to cosmology, seems to make sense. It also explains why our world appears to be at the centre of redshift phenomena. All galaxies appear to move away from us. In fact, this is not necessarily the case. What happens is that the starlight itself, which we see communicates energy loss which becomes larger with the increasing distance of light-emitting cosmic objects.

What are the restraints for a photon with respect to losing energy? Photons can assume a continuum of energy states between very energetic gamma rays and very low energetic radio waves. Energy changes with the wavelength. A travelling photon could lose energy by allowing the wavelength to become longer and by emitting the energy difference in the form of microwaves. Energy in the form of information, which is engaged in the ongoing back conversion of the light wave into a photon particle will be involved. Energy-losing photons therefore just act according to the theorem of dynamically behaving energy. This is quite a remarkable result and could be used as a critical test for the dynamic energy hypothesis. Since there have been so many photons travelling away from stars for long time periods, the reason for most of the radiation of the universe being microwave background radiation can be understood. The energy released from photons accounts for an increase of entropy, and it is to be expected that the microwave distribution will somehow manage to be in equilibrium with the surroundings. Since a larger energy loss, and thus a greater redshift of light, is to be expected the longer the path of travelling photons, such a phenomenon can erroneously suggest a fast expanding universe. When a photon is subject to a measurement, turnover of information will be involved, as we have seen. There will be a kind of "reset" of information with the additional consequence that the light velocity will stay constant. There may also be a "reset" in terms of energy loss or frequency change of the photon. When photons arrive on earth from distant galaxies and are measured, they recreate themselves via the contained information. A constant light velocity would be the consequence, as would be information on the past history of the photons buried in their redshift pattern which reflects energy loss suffered during the long trip. A periodic infrared shift in this case would simply mean a periodically activated mechanism of energy dissipation.

A detailed theory would, of course, be needed for a quantitative understanding of redshift on the basis of the "dynamic energy" approach, but qualitatively the expected redshift can easily be derived from Hubble`s law, which is the empirical law accounting for the redshift of stellar light. Mathematically, this law describes the supposed increase of the escape velocity of galaxies (in kilometres per second) with respect to our earth and is presently the key for understanding the dynamics of the universe including the so-called dark energy:

It describes the galaxy velocity v via the Hubble constant H_o times the distance D of the galaxy from earth in light years (H_o expressed in kilometres per second per megaparsec, which is 3.26 million light years. Its value is estimated to be between 50 and 100, mostly at 70):

$$v = H_o D \tag{15}$$

This formula can, just by mathematical transformation, be converted into a formula expressing the new interpretation of the redshift. If we divide the formula left and right by a distance (kilometres) which is mathematically allowed, we arrive at a statement on the frequency f (1 divided by seconds) of light from galaxies in dependence on their distance from earth D. This way the formula is still valid but allows an entirely different interpretation. The new proportionality factor should now be named F_o and has the dimension of frequency shift per mega-parsec (3.26 million light years):

$$f = F_o D \tag{16}$$

The energy contained in a photon is proportional to the frequency f and a frequency shift per distance (in mega-parsec) consequently reflects energy loss via entropy turnover ΔS. The proportionality factor would be related to the entropy turnover: $F_0 \approx \Delta S$. This means that the shift of light frequency from galaxies changes with their distance D from earth and is dependent on the rate of photon energy loss F_0 via entropy production. The latter, entropy production, is not a constant but would be expected to change logarithmically with the distance from the light source. This simple adapted formula explains, without assuming an inflation of space, why our earth appears to be the centre of space. In fact, it is of course not the centre, but we are seeing from all directions incoming light, which has lost energy on the long journey from galaxies in space. It should be emphasized again that, in contrast to classical quantum theory, the dynamic energy approach allows light to lose energy without interacting with matter, but in response to entropy increase. It simply follows the claimed law that energy aims at decreasing its presence per state. It does this in response to the need of generating entropy during expansion of light into space. This entropy generation through light during expansion into space, which Einstein once stimulated to attribute a particle property to light in analogy to a gas, is presently ignored and sacrificed to quantum physics.

The question of a near or super light velocity expansion within the universe during inflation would lose any basis. Redshifts can no longer be interpreted in terms of relative velocities and inflated space alone. A rapidly expanding universe could turn into a reasonably static one in which far travelling light loses a corresponding amount of energy due to the formation of entropic, chaotic energy. The discussion of dark energy would have to be reassumed from the beginning, since all speculations on such space property were based on light frequency shift studies. My interpretation of entropic energy loss by expanding starlight would, of course, also explain Olbers' paradox. Two centuries ago the German astronomer Wilhelm Olbers recognized that a nightly sky illuminated by a vast space full of stars should always be bright. The reason is that every point of the sky would be occupied by a shining star. That this is not the case has been explained by a fast expanding space and the thereby generated redshift of starlight. The interpretation presented here of travelling starlight, which generates entropy, equally gives a logical explanation and in addition provides an alternative interpretation for the microwave background.

Also for periodic redshifts there would be explanations. There are two ways to understand them. One is by assuming that D in relation (16) changes periodically and the light-emitting galaxies are periodically organized in certain space regions. This is theoretically possible, because the time concept of the dynamic energy hypothesis, the fact that there is a fundamental time flux, allows and favours self-organization of galaxies. The other one would be that F_0 periodically diminishes along the way. This means that there is energy loss in the form of entropy production at reasonably regular intervals during the photons' travel through space. This would be possible because such a property is fundamental in our energy approach. This is in contrast to present concepts, which allow photons only to lose energy when interacting with matter and gravity.

When gravity, which I identified with information related to the particle -wave duality, can self-organize, periodical gravity structures and thus periodical redshifts are also imaginable.

Let us conclude the interpretation of the redshift of starlight on the basis of the dynamic energy approach: it can apparently be explained without further assumptions and without assuming an exploding or inflating space.

23 The Big Bang irrationality

The redshift of characteristic light emission from stars, increasing with their distance from our solar system, has become the most discussed argument for an explosive start of our universe in terms of a Big Bang. Before, there was no space and time with energy and matter starting from a timeless past in form of an extremely condensed and superheated nucleus. The formalism of the general theory of relativity can be adapted to support this kind of cosmic evolution. As mentioned, the Standard Model of the Big Bang Theory describes many steps of space evolution starting from a time of 10 to the power of -43 seconds (one second divided by one with 43 zeros). This presently shortest imaginable time is the so-called Planck time which depends on natural constants only, the velocity of light and the gravitation constant. The inflation, the explosive increase of the seed universe, is, for example, assumed to have occurred between 10 to the power of -33 and -32 seconds. However, it is also known from the general theory of relativity that the time should slow down with increasing gravitation. How great was gravitation around the Big Bang seed, when most energy-mass of the universe came from it? Does a concept of time still make sense? There is another irritating fact: matter and energy distributed into a vast space generates a great deal of entropy in the form of non-usable energy. Where is it located? The microwave background radiation has, in fact, been interpreted as arising from "stretched" and red-shifted light from the Big Bang. Where is the entropy, which was generated in the cold universe by expanding energy and matter over a period of more than 13 billion years? How can the time-invertible physical laws, introduced into general relativity theory, explain an exploding universe at all? I do not consider this elaborate Big Bang scenario rational.

The uniform microwave background radiation, which I understand in a different way, is considered to be the second most convincing support for an evolution of the universe from a Big Bang explosion. In the extremely hot plasma, quantum fluctuations and spontaneous appearances of particle-antiparticle pairs are expected to have occurred. They have not really existed for a long time, but are considered to be "virtual" particles. Nevertheless, it is claimed that their existence does not violate energy conservation laws. They can apparently be the origin of causality (which is claimed here, while it is denied in other situations). They are supposed to have induced small accumulations of matter and energy, which later resulted in galaxies and clusters of galaxies all over the universe. Small variations in the temperature of microwave background should reflect such an evolution (fig. 30). The satellite COBE measured them and found, distributed over space, a deviation from the average temperature amounting to only 30 micro-Kelvin. This means that the temperature, which microwaves yield, varies by 30 millionths of one degree Kelvin (or Celsius). The discovery earned J.C. Mather and G.F. Smoot a Nobel Prize. Today a series of theories have been elaborated to explain details of these extremely small microwave ripples.

However, to reach an explanation for the quite evenly distributed microwave background in space, the extremely small Big Bang seed had to undergo an enormous amplification intended to allow an even distribution of energy and temperature in space. The general theory of relativity already appears to describe the Big Bang if one constant, the "cosmological" constant", is set at zero or close to zero. In 1981 the particle physicist Alain Guth found that assuming other values of the cosmological constant, meant the general theory of relativity permitted an exponential increase of space. It became known under the name "inflation" and is supposed to have occurred within 10^{-32} seconds. This is one second divided by a one with 32 zeros. Space should have increased by a factor of 10^{30}. This is a one followed by 30 zeros. Einstein once said that "time is what the clock shows". What clocks showed the time during inflation? And why can theoreticians predict such fast processes from a theory with time-invertible laws and without considering the unknown intensity of gravitation? In addition, Einstein's general theory of relativity does not respect the conservation of energy. Energy could be created from nothing. Is such a theory reliable? Today one already knows many variations of the inflation theory, which, depending on the assumptions and parameters involved, can describe nearly any experimental observation. An even greater problem seems to be that under given mathematical conditions, multiple universes of space-time inevitably tend to occur and to continue multiplying. They could explain any desired property of a universe. Energy may disappear in one universe and show up in another. One has the impression that these are convenient tricks to avoid restrictive natural laws. The inflation theory, or a chaotically reacting mathematical singularity, turned out to be a helpful tool for tracing down space mysteries. Inflation theory has, for example, also been used to explain why the universe contains so much matter (around 500 billion galaxies). Quantum mechanics, via the uncertainty relation, permits energy fluctuations from nothing. These fluctuations may generate particles and anti-particles. The vast expansion of space during inflation is thus associated with the quantum mechanical creation of this matter. I recall that in chapter 14 I argued that there is no reason to interpret the uncertainty condition that way. Generation of matter and energy from nothing is quantum fiction.

Within the Big Bang theory, questions on energy balance are answered in the following way: "energy contained in this matter, and virtual (negative) energy contained in gravity just balance. The universe contains zero energy" (Hawking, 1998). Can zero energy produce all the dynamics of the universe? And what about dark energy and the explosive expansion of the universe?

The postulated inflation of the universe also offered an additional opportunity for imaginative scientists. Since the hot Big Bang origin of the universe implies the primordial existence of a local thermal equilibrium, which is not a good start for the building up of order, a rapidly expanding universe opens an escape. In a vastly expanded universe the maximum permitted entropy, disorder, would become much larger. This permits a development of the newly formed universe towards increasing entropy (Penrose, 1989). The irrational inflation of empty space, therefore, not only allows the filling up of the universe with matter and popping up from nothing, but also the development of local order via entropy production. Can this happen on the basis of time-invertible basic laws?

On March 17th, 2014 physicists working with the PICEP2 experiment on the South Pole announced the detection of remnants of gravitational waves in the microwave background radiation. What they in fact measured was a slight, apparent periodic spatial optical polarization in the microwave background radiation studied at a selected 2% section of visible outer space (select 2% from fig. 30). Their conclusion was that they were seeing gravitational waves related to the Big Bang explosion and inflation of the universe. Such waves, which the general theory of relativity permits, could not yet be measured. Gravitational waves are periodic compressions and expansions of space-time deducible from the general theory of relativity. Within such waves, time is expected to oscillate. This is, as I already mentioned, an incredibly unrealistic phenomenon. Is this what we understand when talking about clock-time? Most newspapers and many Internet sites were reporting the cited claims that this is a definitive proof of the Big Bang, and of inflation theory. One day after the announcement, and before even scientific publications appeared and critics could be heard, the hailed big discovery and breakthrough was already immortalized in Wikipedia. This is the impatient, aggressive and manipulative way some areas of science work today. They make bold statements and claims without even waiting for comments and critical questions of referees. Repeatedly the opinion was expressed that this discovery deserves the Nobel Prize. Soon, however, the science community agreed that the measurement simply analysed stardust.

Since we are dealing with the apparent confirmation of mathematical models it is interesting here to also cite comments Einstein himself made about mathematics in relation to reality; " insofar as the theorems of mathematics relate to the truth, they are not certain. And insofar as they are certain, they do not relate to the truth. Mathematical theories in reality are always uncertain. When they are secured, they do not reflect reality". Einstein also said: " Since mathematicians took charge of the theory of relativity, I myself do not understand it any more" (Mathematik-Zitate, 2014). These comments were probably too pessimistic, since mathematics has proved to be an indispensable tool for physics. And there are, of course, entirely contrasting comments, like those from Nobel Prize winner Eugene Wigner who wrote an essay on "The unreasonable effectiveness of mathematics in natural sciences".

Nevertheless, there are some critical aspects. One is that there are sometimes several or many solutions of an equation available to describe a natural phenomenon. Only that which comes close to reality is typically selected,. The theory then is valuated as proven. What about the other solutions? Complicated matters often attract less attention by theoreticians than simpler ones. Even though everything changes in nature, natural laws are practically all formulated to be time-invertible. This means that changing the time direction does not change the law. Reality, however, develops only in one direction. Time-oriented processes like self-organized mechanisms are not easily explained. This is not satisfactory. From the spiral-shaped elegance of far galaxies, to biological evolution and to geological features such as meandering rivers, everything originates from such mechanisms. Another point of criticism is even more serious and I have already mentioned it in relation to our classical statistical definition of time orientation (chapter 8). When complicated mathematical relations describing a natural situation are simplified, information is taken out from them. Information, however, has an energy content, so this procedure will manipulate the described system. This is usually not considered when applying formulas. But formulas which few people really understand today, and non-intuitive quantum mechanical mechanism are actually shaping present concepts of cosmology.

How does the dynamic energy hypothesis presented here see the Big Bang hypothesis?

First, a four-dimensional space-time is not required, because the experimentally verified constant light velocity is explained locally. As a consequence, the general theory of relativity should not be applied to simulate a Big Bang explosion. It follows that inflation of empty space cannot be invoked as a mathematical possibility. A redshift increasing with the distance of galaxies is explained as reflecting energy loss and entropy formation by photon fluxes expanding into space. Such an energy loss is possible, because it reflects the basic assumption of the proposed dynamic energy hypothesis, and because a fundamental directional timeflow is active. Therefore, an explosive expansion of the universe backed by a corresponding redshift is not at all evident. In addition, the microwave background in the universe is explained as the entropic- "chaotic" energy lost from starlight photons, which is now seen distributed in space. The celebrated recent discovery of traces from gravitational waves in the polarized patterns of the microwave background radiation claimed to prove the Big Bang and space inflation and can be judged in a similar way. First, the new energy model allows energy to self-organize. Pattern formation may therefore also occur with the microwave background radiation or stellar dust. To claim that the half dozen, extremely weak and barely periodical patterns within a 2% space section reflect gravitational waves is very audacious. Several well-designed experiments have up to now not been able to verify gravitational waves, because they either did not show up or were too weak. In the microwave background measurement the wavelengths of the supposed gravitational waves observed, covering a 2% space section, must amount to millions of light years. Such long waves have never before been discussed in science.. One evident reason is that the energy contained in waves decreases inversely proportional to the wave length. A very long wave length reflects a very low energy content. All together it does not make sense either to speak of the meaning of a periodically changing time flow expected to occur in a gravitation wave.

The irrationalities associated with the Big Bang scenario are very significant and could be avoided with more reasonable assumptions about the universe. A fireball at the beginning actually means maximum disorder, a maximum entropy situation. However, a low disorder, low entropy situation is needed to explain the evolution of structures and patterns in the universe which are actually seen. Various mechanisms have been taken into consideration to explain this increase of order after the Big Bang (Penrose, 1989) (Layzer, 1975, 1977, 1988, 1990): quantum indeterminism, energy and matter popping out from nothing when space expands, is an imagined source of order and drive for the universe. Matter and energy are then created within an expanding space.

After identifying quantum indeterminism as a chaotic determinism arising from self- organized energy-matter, this interpretation is excluded. Energy and matter popping out from nothing is, within our understanding of energy, an illusion (chapter 14). Other concepts argue that in an expanding universe the permitted maximum entropy, the number of permitted states, is rapidly increasing and exceeding the entropy of the original tiny Big Bang seed (Layzer, 1990). Here one is forced to believe that empty space can really expand and that the energy needed is arriving from somewhere. Also the still poorly elucidated properties of gravitation were invoked for the explanation of order arising from the Big Bang scenario. We interpreted gravitation as an information self-image related to the particle -wave dualism. It may also be considered acting as a self-organized mechanism, which is active towards the generation of order via a superior form of gravitation. This will be discussed later.

A Big Bang origin of the universe is not an acceptable option for the dynamic energy approach as it contains too many irrationalities.

24 Cosmological objects from galaxies to black holes

The attempt to describe nature rationally developed here understands matter and elementary particles as self-organized energy. This is possible, because energy conversion drives time. The timeflow is thus fundamental and not a statistical phenomenon. There is accordingly a before and an after in basic processes. Gravity itself has been identified as a self-image of matter in the form of information. It has an energy content and mediates between the particle and wave property of matter. This information therefore also involves the claimed fundamental drive of energy to decrease its presence per state. By attracting matter, gravity indeed decreases free energy and generates "chaotic", entropic energy. Classical physics talks about potential energy which can be gained from a gravitation field. Water from a high-positioned lake can be used to drive an electric power station. This way energy can be extracted from gravitation. As a parallel product entropic energy is generated, for example in the form of heat, vibrations and rotation of atoms or molecules. This shows that where gravitation is active, energy is present in free space and tends to decrease the energy per state. This is the basic property of energy claimed within the discussed new approach. Besides, matter and information, the latter was identified with gravitation, act together as a self-organized energy system towards that aim. Fig. 31 shows how gravitation interferes with the self-organized spiral patterns of galaxies at a very large distance. Here a super-gravity may be involved. It could be the consequence of self-organization of gravity, which itself was identified with the information self-image of matter (chapter 21 and 28). The diameter of a galaxy may range from a hundred thousand (our galaxy) to one million light years.

The formation of elementary particles via dynamic self-organization of energy has already been discussed. In the same way elements could have formed which make up the periodic table. Many of those require a certain degree of order to accommodate the required composition of protons, neutrons and electrons. Synthesizing elements far from equilibrium by building up order at the expense of entropy could be a much better working strategy than just allowing the collision of the building stones of elements at high temperature as is presently assumed. Today, element formation in the universe is entirely explained via the Big Bang, during which mostly hydrogen and helium is supposed to have been formed, and much later the consequences of star explosions apparently led to elements with higher masses. While element formation under extreme conditions of exploding stars cannot be ignored, dynamic self-organization in energy-rich, high temperature environments equally deserves being investigated. The prospect of being able to build up order in an otherwise chaotic environment which operates far from equilibrium is an attractive opportunity for understanding synthesis of heavier elements.

Fig 31. These two interacting galaxies (NGC 5257) visualize the action of gravitation in free space over long distances (NGC, 2014, credit: NASA). Gravitation acts very far and penetrates matter. It is here claimed to be a phenomenon of information, a self-image of matter, controlling properties of matter. Information can also self-organize, generating a higher hierarchy of information, a kind of super-gravitation, which keeps together the self-organized, rotating spiral arms of the galaxies.

The advantage of self-organization for chemical synthesis is well known. Bacteria fix nitrogen in our environment with a genetically controlled catalytic process at ambient temperatures. Local order is used here to greatly favour energy requiring chemical mechanisms. Technically, high temperatures of at least 400°C and pressures of 200 atmospheres are needed for this important process as are selected catalysts. Our knowledge of nuclear chemical processes is too modest to understand mechanisms during which order may locally build up in support of a much improved catalysis of element formation. However, in a time-oriented universe there is no reason why self-organized element formation should not exist and did not occur during the evolution of the universe.

Few space objects have stimulated the imagination more than black holes. They concentrate mass and gravitation to such an extent that nothing, not even light, is expected to escape its boundaries. They will therefore essentially have a black appearance. First speculations about black stars date back to 1783, and today more than a dozen candidates are suspected in our galaxy alone. Their estimated size ranges from several solar masses to a black hole of 4,3 million solar masses. The latter, Sagittarius A*, forms the centre of our galaxy (Sagittarius A*, 2014). Most galaxies are suspected to have black holes, which attract and devour stellar dust and stellar masses, in their centre.

There are different strategies available to identify black holes. They range from observing neighbouring stellar objects to detecting gravitational lens effects, a kind of mirage phenomenon produced by gravitation. A further characteristic property is apparently rays of charged particles and of radiation emitted perpendicular to the plain of a rotating black hole.

Today black holes are essentially considered to be phenomena subject to the theory of general relativity and to quantum theory. Both are in fact equilibrium theories with a time which is either an illusion or absent. When certain critical masses are exceeded, a gravitational collapse occurs and space-time is expected to be dramatically curved and concentrated to such an extent that a kind of singularity is generated. Small parameter changes can have dramatic consequences. This is a reason why little can safely be concluded about the function of a black hole. Imaginative quantum mechanisms have, however, been invoked to escape from this situation. Energy can pop up from nothing generating virtual particles and anti-particles. The latter may fall into the black hole allowing the counter particles to escape gravitation. This is the famous Hawking radiation from black holes which gradually eats the same away. The radiation is however predicted to be so weak that it will not be measurable. Also thermodynamic approaches are attempted in black hole theory, even though a function according to equilibrium thermodynamics is not obvious. It has to be pointed out again that both the general theory of relativity and quantum theory build on fundamental laws which are time invertible and describe reversible processes at equilibrium. Nevertheless theoreticians have the courage to use them for describing a highly irreversible, " matter eating" phenomenon such as a black hole.

Are these convincing descriptions of the black hole function? What could be said about black holes from the viewpoint of the "dynamic energy" approach? Here the simple basic postulate that the energy tends to decrease its presence per state has a straightforward consequence: a far from equilibrium situation and a maximum entropy production within the constraints of the system.

From such viewpoints alone, it should be clear that black holes must then also be phenomena of self-organization of matter and energy. But where does the energy go in a black hole? This energy has to be conserved, but is no longer useful and is the product of entropy production. Where does it go when not even light is expected to escape from a black hole?

Black holes are typically found in the centre of spiral galaxies. Superficially they fit into a pattern seen with hurricanes and typhoons. They thus exhibit structures typical for pattern formation during energy turnover far from equilibrium. This is possible because the fundamental claim of the "dynamic energy" hypothesis involves a time arrow, an energy property related to change. In this picture, a black hole is something like a vegetarian dinosaur feeding on and devouring vegetation in a Triassic forest. Gravity would play a role like the food harvesting activities of the dinosaur (fig. 32). Matter disappears into the black hole like leaves into the mouth of the dinosaur. Looking at such a picture what can then be said about entropy production by a black hole? According to what has been concluded before, a definite trend towards maximum entropy production is to be expected (chapter 11). But where are the constraints of the system? Where does the "chaotic" energy in the form of entropy go when not even light can escape a black hole? What does a dinosaur do when his digestion system cannot get rid of unusable products? He either gets seriously sick and dies or he must get rid of his excrements. In irreversible thermodynamics the self-organized system would, in such a situation, shift further away from equilibrium and new parameter constellations would start dominating within the constraints of the system. The physical situation in a black hole would be characterized by the increasing intensity of gravitation. Gravitation was identified with the information self-image of matter, aimed at decreasing the presence of energy per state. With increasing gravitation there will be an increasing drive to get rid of "chaotic" energy in the form of entropy. Modified or new self-organization patterns with new properties would show up, for example an association of a black hole with a quasar. However, also self-organization of gravitation may be possible, a kind of super-gravitation. Here information, which gravitation represents, may build up a kind of order - a hierarchy above pure information. Much higher gravitation is to be expected, as are entirely new additional properties. As an example, information processing in our brain should be mentioned. If self-organized information handling is to enable consciousness, a sufficient energy flow has to be maintained through the system. What self-organized gravitation in a black hole will achieve can at present only be speculated.

a)

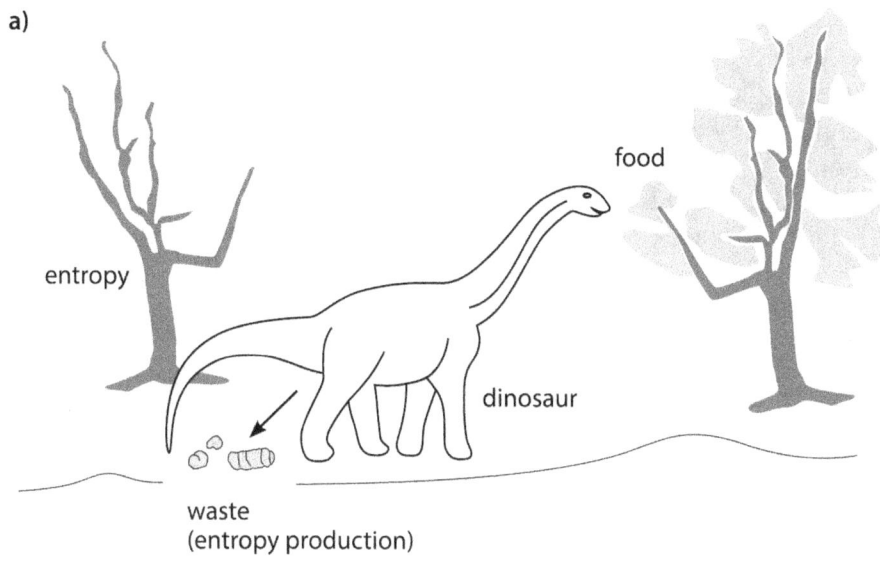

food

entropy

dinosaur

waste
(entropy production)

b)

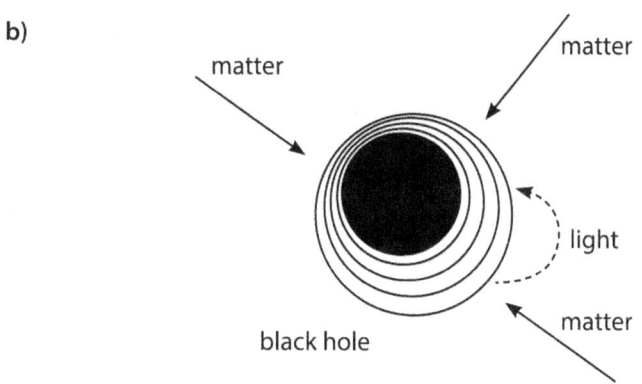

matter

matter

light

matter

black hole

Where does the produced entropy remain?

Fig. 32. A black hole may be compared with a vegetarian dinosaur feeding in a Triassic forest. Both systems are controlled by self-organization, but the black hole has to find a way to get rid of entropy, degraded energy.

For a developing black hole specialized self-organized structures are imaginable - structures which finally get rid of degraded energy, for example in the form of radiation. Quantum concepts cannot rationally deal with such dynamics. In contrast, within a time-oriented energy, a black hole has to be understood as an inorganic product of dynamic self-organization with the ability to react to parameter changes and to energy flow. It is known today that such phenomena need time orientation, far from equilibrium conditions, and feedback processes. Neither the general theory of relativity nor quantum theory can provide such conditions. The Schwartzschild solution of the general relativity equations from

1916, which generated the black hole prototype and still serves as reference, has nothing to do with dynamic self-organization. It just describes an imagined increasing deformation of space and time around an accumulating and contracting mass. I doubt that empty space can develop such intricate behaviour as "bending" space and time. It is recalled that "bending" time in space-time means manipulating the flow of time. I do not believe it is this time which really matters for black holes.

Space is full of beautifully structured galactic objects and dynamic self-organization appears to be present everywhere. This seems to be a significant factor for creativity in the universe and entirely relies on time-oriented mechanisms. Allowing for a fundamental time orientation, as I did with my energy proposal, will also allow dust and matter in a highly diluted form to develop complicated patterns. But there must, of course, also be an energetic drive that pushes the systems far from equilibrium.

Quasars are especially interesting objects and are typically associated with a black hole in the centre of galaxies with a bulge. This combination could explain how a black hole dinosaur finally got rid of his excrement. The black holes collect matter and star fragments and rotate them in a so-called accretion disc. From there matter can fall into the black hole or be radiated away with total intensity amounting to a million to a billion times our solar radiation. In fact, radiation intensities equivalent to that of one hundred Milky Way galaxies have been estimated. Such radiation emitting systems are called quasars. They belong to the most brilliant objects in the universe and may originate from black holes which have drifted further away from equilibrium and have developed better adapted new structures for energy turnover. And, since they appear to be associated with black holes to which they donate matter and from which they subtract energy in the form of radiation and dust, they testify for maximum entropy production during conversion of solid matter into liberated waste energy. In fact, nuclear experience on earth can help us to understand such a process. When matter is turned over in nuclear or hydrogen bombs, one starts from nuclear material with a specific weight and very limited radioactive energy turnover. After a nuclear explosion during which matter is converted into energy, and after the dramatic energetic changes that accompany that process, the high temperature, high radiation and radioactivity gradually fade away. Maximum entropy or disorder, non-useful energy, will remain in the environment. This could well be monitored on nuclear testing sites from a longer period of time ago.
Such processes of turnover of matter subject to maximum entropy production are also expected to occur around quasars in combination with black holes, with the difference that enormous quantities of matter are consumed. The energetic "evaporation" of matter from quasar -black hole aggregates via nuclear processes aims at generating a maximum distribution of "chaotic" energy in space and time. They are basically executing, to a maximum performance because of optimized

constraints, what I claim to be a fundamental law: energy's tendency to decrease its presence per state in space and time. The quasar black hole assemblies seem to have reached an evolution state where maximum energy turnover and, correspondingly maximum entropy turnover, is approached by a self-organized system. A black hole alone would, because of extremely high gravity, have difficulties in emitting radiation and matter. But evolution has apparently found a way out by modifying it and turning it into an association with a quasar. The result, a self-organized system with much improved ability for entropy production, is convincing. This may support my interpretation of an "aim" in evolution (chapter 27) which is maximum energy turnover. In addition, it may testify for a long and dynamic evolution history of galaxies in space. Quasar-black hole associations also change in light intensity within weeks or months. They are inorganic systems essentially behaving like living animals. I compared them with vegetarian dinosaurs. Their activities depend on the existing constraints. Dinosaurs also generated less entropy when finding and digesting less food.

The interpretation given here for black holes is, of course, drastically different from present interpretations in which equilibrium thermodynamics and quantum theory is adapted to fit with the space bending and event horizon concepts of general theory of relativity (Black hole, 2014). Also, the role of the quasar is interpreted differently: it is just a secondary phenomenon. Its enormous light emission results from the accretion disc, where matter rotates while waiting to be accepted by the black hole. Energy is now essentially thought to be liberated through friction. In contrast, our interpretation is that evolution of an inorganic self-organized quasar-black hole system has optimized energy turnover for maximum entropy production within the constraints of the system. When contracting matter develops towards a small black hole, energy turnover is working well, but with a black hole growing, chaotic energy in the form of entropy, small particles and radiation can no longer leave the system. The black hole will temporarily serve as a grave for entropy, because gravitation is too high. This, however, would affect the distance of the system from equilibrium. The self-organized system consequently restructures and "searches" for a way to get rid of entropy. This search is simply the consequence of the recognized fundamental energy property. Entropy confined into a limited space has a higher value of energy per state. Larger black holes finally find a way to get rid of energy in the form of entropy, chaotic waste energy. They do it by restructuring in such a way that entropy, non-useful energy, can be ejected. This apparently happens via the so-called accretion disc which forms around the black hole and handles the transport of matter-energy to the black hole and of entropy, non-useful matter and energy, from the black hole (fig. 33). Also self-organization of gravitation (information, see chapter 28) may occur and support this process of maximum entropy liberation. That the accretion disc is a product of dynamic self-organization is not only evident from its spiral structure, but also from quasi-periodic oscillations which have been observed to occur in these. Between 20,000 to 100,000 quasars have already been discovered to occupy the centres of galaxies. They have the size of our solar system, but can radiate more intensively than 100 galaxies together. The radiation energy is very widely distributed from gamma rays to ultraviolet light, from visible light to heat radiation and microwaves. This is in contrast to the radiation from galaxies, which mostly radiates in the visible and infrared region. Due to the combination with quasar activity, black holes have apparently found a way to get rid of matter and energy generated as confined entropy and to approach a condition of maximum entropy production. The difficulty of getting rid of entropy from a black hole may have shifted its activity further away from equilibrium, exploring new structures and finally allowing entropy to escape from gravitation. Self-organization of gravitation, super-gravitation, may enable a spatial structuring of gravitation, allowing for gravitation free channels, which facilitate the emission of beams of radiation and gases into free space.

Quasars were found to be abundant 2 to 4 billion years after the start of our universe, much less abundant before and after this time period. This information on the time window for quasars should be reinvestigated, since the presented new approach offers a different interpretation for the cosmic redshift of light. However, it could be that quasars became only active during a certain period of evolution, like vegetarian dinosaurs in their Triassic forest environment.

maximum entropy production

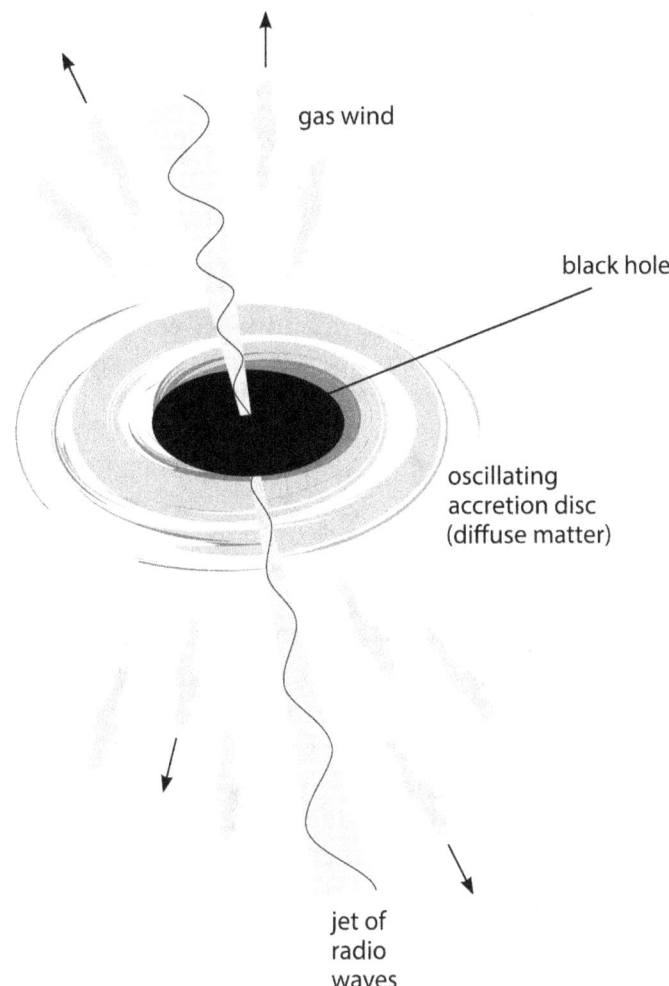

gas wind

black hole

oscillating
accretion disc
(diffuse matter)

jet of
radio
waves

Fig. 33. Association of black hole and quasar during maximum entropy production. Both are dynamically self-organized systems, which, depending on the energy turned over, vary their distance from equilibrium and their parameter constellations. Maximum entropy emission occurs through the accretion disc, which appears to serve as a logistic platform for the collection of useful matter and emission of waste energy. Self-organized gravitation (information) may be part of the mechanism.

TIME, EVOLUTION AND MIND

25 An alternative brief history of time

As long as people have tried to understand the world, they must have speculated about what time really is. Since antiquity many intelligent comments about time have been circulated in philosophical and literary environments. During the last two centuries numerous books have also been written about the meaning of time. I myself, when exploring the "dynamic energy" idea, evaluated the conventional concepts of time (Tributsch 2008) in order to learn what meaning they would have for me. I selected the title of this chapter as a contrast to Steven Hawking`s book "A brief history of time", because nearly all conclusions which I reached are different.

Science today uses time just as a ruler to monitor changes. All basic laws are postulated to be time-invertible. A time orientation in the direction of increasing disorder which can be witnessed by everyone is just understood to be a consequence of statistical processes. Microscopic uncertainty and chaotic mechanisms apparently allow changes only in one direction, they do not allow a return. Since the rise of the theory of relativity, as promoted by Einstein and to some extent by Poincaré, which abandons an absolute time and emphasizes that time is relative, physicists tell us, that time is actually an illusion. Einstein himself wrote in 1955 in a letter to the son and the sister of his deceased friend Michele Besso: "..the distinction between past, presence and future has only the meaning of an illusion, though a persistent one". For physicists, this statement did not sound too unfamiliar, considering the fact that all natural laws were anyhow formulated to be time invertible. No need for a fundamental time arrow (as I am introducing it) was recognized. The natural laws which physics teaches, could also accompany our universe into the past, but our universe does not like that. However, Einstein`s relativity theory, in fact, tells us only that time is not an absolute quantity, but is relative depending on relative movements and on its complex unity with space in a four-dimensional space-time. Every object and every individual person has a proper time depending on position and relative velocity. Therefore, the statement "time is an illusion" is more equivalent to asking "what time should one take"? Since this cannot be answered, scientists agree that formulas describing our world should avoid expressing time. De facto this means that we should forget about time. It is just "what the clock shows" (a comment by Einstein), a periodical phenomenon which is calibrated to monitor the progress of change.

Let us now think a little bit about the meaning of time in our "dynamic energy" world.

We will again look at cognition from the viewpoint of Ernst Cassirer's theory of "symbolic forms". In the new picture of an energy, which has dynamic properties according to statement (7) and relation (11), an entirely new interpretation is possible. Fact is that there is no change without energy conversion. A timeflow is recognized in the direction of generated changes (which also corresponds to the direction of entropy increase (increase of disorder)), because entropy production is a consequence of energy conversion. A timeflow into the past is therefore not possible. Energy conversion and its "chaotic" products, which lead to entropy production, would generate the timeflow only in the direction we experience. This is easy to understand, since it is the changes around us and within our body, which we observe and which give us the impression of timeflow. They determine the front line of what we recognize as passing time. Changes that happened in the past define the past, present changes define what happens now, and expected future changes define the future. Here, time is not an illusion. Time which has passed can be characterized in terms of energy which has been turned over. Present time is paralleled by ongoing energy consuming processes. Time in the future is expected to turn over energy later and there is an important fact: the interrelation of a time-creating energy with time itself specifically means that "time-oriented" energy and time depend on each other. They are interlinked, which also means that energy turnover generates "time", but not in the way we define time, but in the form of action (energy multiplied by time). Action, as we have seen, plays a fundamental role in physics. Many natural phenomena and physical laws follow the principle of least action. The chains of elements of action generated by ongoing energy conversion, reflect the reality of "time-flow". Earlier I showed that both oriented-energy and oriented -time can be derived from the dynamic principle of least action (chapter 10). When formulating this principle as a sequence of infinitesimally small integrals over energy multiplied by time one realizes that they can be minimized only if energy and time have oriented, vectorial, properties. A minimisation cannot be performed if they are considered to be mere numbers, "skalar" quantities as they are dealt with today in physics. Individual numbers have no tendency to minimise themselves, but general energy and time quantities can have. The principle of least action is, in fact, a different statement of the "dynamic energy" hypothesis. It is a dynamic principle, which had been forced to function in a time-invertible world.

It should therefore be repeated that within the dynamic energy approach, the generation of action, energy multiplied by time, is the only fundamental time-related phenomenon in nature. It is claimed here that time in its "pure" form, liberated from energy as we derive it from our clocks, does not exist in nature. Only those phenomena exist which can be directly measured. Time cannot be measured like a current, like the wind or water flow. For a direct measurement it would have to be associated with matter or energy. Only when combined with energy, as action, is time measurable and our clocks basically measure it that way. Sequences of action, and not time intervals themselves, are consequently the quantities which are expected in nature to generate our experience of time. Time has no substance and there is no mechanism to monitor time generation in nature besides clocks, which read time out of action. Time should always appear buried in chains of action and has to be extracted from chains of elements of action by dividing and thus calibrating them by the energy turned over. This is done by clocks, in the brain or in plant seeds or organisms that depend on time programs. Time in its pure form is something artificial, something extracted from the path of action which energy conversion draws into our environment. Nature, by the way, does not aim at perfect clocks. Our heartbeat, for example, is not exactly periodic even at rest. The time series follows a fractal, self-similar behaviour indicating a certain degree of uncorrelated randomness. Disease, sleep and age can affect the fractal behaviour. Fractal behaviour indicates that self-organized mechanisms are involved. Such mechanisms are based on energy flow. This confirms that time in nature is extracted from irreversible phenomena. An irreversible phenomenon of this kind is friction, which is occurring when a moving object is interacting with another medium, a solid or a liquid. Already when just a few molecular vibrations, thermal products of friction, are produced, the process cannot any more be reversed in time. And this is not yet a statistical process.

Let us return to the statement of the theory of relativity that every observer in a moving reference frame has his own time. When using light to measure the time of a moving object from three observation points shifting at different speeds, three different time recordings will result (fig. 26). What then is the real time on the registered object? According to the theory of relativity, the time measured concerning the three moving objects will pass at a different rate according to relation (14). I say that real time works differently. The three observers should measure the generation of action (energy multiplied by time). Even within the theory of relativity, action is invariant in different moving reference systems. All three observers will measure the same flow of action. Then the observer on the moving object should be asked to measure or estimate the energy turned over during the flow of action. The result, after dividing by energy, will be the chain of time intervals passed. This is expected to be the time which energy-converting people are really experiencing on the moving object. It should also be identical for the observers on the reference frames with different velocities if they divide action by the energy they measure in their system. Life will proceed equally fast on systems moving with different relative velocity.

If, in contrast, on the basis of relativity theory clocks are monitored via light signals from an observing station, time dilation will be observed. It is caused by the finite light velocity which distorts information. The dilated time is not equivalent to the time extracted locally from the flow of action. Action, as pointed out, is invariant against transformation to other reference systems with different velocities. This means, as we have seen before, that time travel for energy-converting systems such as living beings, with all the dramatic contradictions to thermodynamics and the practical experience of life, can therefore not be a real phenomenon even within a re-interpreted theory of relativity. Travelling back in time to change the path of destiny is an illusion. Additional paradoxes and irrationalities which would be possible via time travel are consequently eliminated.

For energy-converting systems time is not an illusion (due to being relative) because they experience time as action, energy multiplied by time. Many living beings have a feeling for time. I explain this in such a way that organisms somehow measure the flow of action and divide it by some physiological activity which is related to energy consumption. By dividing the flow of action by an energy flow, they may get a feeling for time. For us this will happen in the brain. Periodical chemical processes which are known to occur in organisms may support that, providing technical conditions similar to those with clocks. Of course, physiological parameters may change with the living species and with their health. As a subjective experience time indeed proceeds at a different rate for different living species. Also within the same species the rate may change depending on given conditions. For older people time appears to proceed faster. Also illness and fever may change the felt flow of time.

When quantum states are generated, when energy conversion is suppressed, no timeflow is activated. But what time is time in the theory of relativity? It is an artificial time separated from energy and action. Quantum phenomena, which do not convert energy, are expected to follow relativistic properties subject to the special theory of relativity, because for quantum systems time has lost its relation to energy. Independent transformation of energy and time is possible, yielding time dilation (e. g. modifying the energy and lifetime of a meson, or affecting time in travelling atomic clocks). Here time becomes a mathematical parameter invented by man. And what are clocks which measure time? All measured time is a measurement of reference points typically based on periodically occurring phenomenon. This can for example be the rise of the sun or the periodicity of the longest day of the year. And energy is needed to measure this periodicity. When looking at the periodicity of the moon, we activate information to register the arrival and position of the moon. And this again involves energy. Periodicity, measured by an energy-driven tool, generates an artificial time which is not a lived reality. It is a creation of our technical imagination or a situation generated in our brain. It was created to simulate the progress of action around us. Reality is not this artificial time but the flow of action.

The theory of relativity uses clock time and allows transformation of time and energy independently. It thereby creates a partially artificial world. I say partially because phenomena during which energy does not create time in the form of action, such as quantum phenomena, fit reasonably well into the theoretical expectations. This may explain why predicted relativity phenomena are experimentally observed. But an ambiguity remains. Why should time dilation disappear (formula (14)) when a hypothetical ideal signal transport approaching infinite velocity ($c \to \infty$) is assumed? This suggests that the relative "slowness" of light velocity c with respect to the velocity of the measured object, generates the strange phenomena of relativity. Explained in a simplified way: for an object moving very fast away from an observer, the rear edge will appear closer, because light needs time to cross the distance across the object.

Such considerations show how the choice and handling of symbolic forms can either generate irrationality or rationality. Steven Hawking, for example, considered it possible that time in space-time could form a closed loop and continue forever. He went so far as to say that in this case a creating God would not be necessary (Hawking, 1991). According to the arguments presented here and also according to the 2nd law of thermodynamics, this is not possible. Closed systems, and the universe was considered as one of these, increase their entropy, accumulating waste energy. From within, the system cannot return to the initial situation. Only when the system is left open can information from outside be used for this aim. This latter possibility is considered in this study (fig. 37).

Another example is time travel with all its potential for irrationality. The relativity theory allows it (Einstein 1911). According to the presented dynamic energy approach, it is impossible. The reason is simple. An energy-consuming space traveller will have to travel with action, energy multiplied by time, and not with time alone. And action does not care about relative velocities of reference systems. It remains invariant. We have arrived at a "time" concept which is rational. Time is definitively not an illusion as science presently claims, but fundamental when seen as linked with energy. Sometimes I have the impression that ancient religions have intuitively reached a clearer concept of what time is (in relation to energy) than science. In the Tibetan Kalachakra - Tantra, a sacred text which contains the Shambala myth, one reads: "all beings are created by time and destroyed by time" and "movement generates time". Indeed, movement, or change, generates time via a generation of action, and action creates and terminates life.

26   The time arrow and self-organization of matter

As already mentioned, a time arrow allows matter to self organize, provided that feedback processes are acting. The property of a directional timeflow is mathematically and physically required because feedback processes are not possible without a "before" and an "after". Classical physics has the problem that all fundamental natural mechanisms are defined to be time invertible. Time-oriented processes, as readily seen in our environment, and self-organization of matter, as visible in inorganic and organic phenomena including that of life, should therefore not be possible. Science has therefore put much emphasis in mathematically deriving and explaining time-oriented behaviour from time-invertible basic laws. As explained previously (chapter 8), the statistical time arrow in the direction of an entropy increase was the result. It was emphasized that it followed from the mathematical trick of abandoning information. This is not at all justified because this changes the energy content of the system and that way, of course, interrupts the way back in time to the original situation. When Boltzmann made the first advance in this direction, he did not know that information has an energy content. He eliminated information from his statistical formula and thus eliminated energy. This way he changed the quality of the system. It could not any more return to the initial state. It became "oriented", as he wanted it to become. But phenomena in nature are really oriented. Already a small movement of an object, which is exposed to friction is oriented in time. The tiny heat vibrations, the entropy generated, simply make the difference between past and future.

When models of self-organization processes began to be accessible to computer-calculations, it was shown that entropy export and generation of order become possible beyond a so-called "bifurcation point" when the system begins to assume new dynamics. The Russian-Belgian scientist I. Prigogine and his school were quite active in this field, but the mathematically observed self-organization remains an illusion. Only when feedback processes work, can self-organization develop. This can be mathematically demonstrated. However, time-invertible statistical processes cannot provide adequate conditions for feedback. All mathematical processes invoked to introduce such conditions include an abandonment of information and thus of energy. Here mathematics has been used to manipulate physical-chemical systems.

For the "dynamic energy" approach, such a problem does not exist. It states that there is a fundamental time arrow. It naturally proceeds in the direction of energy turnover and entropy production though by turning over elements of action, or energy multiplied by time. The properties defined are exactly those which make such self-organization mechanisms possible. The "symbolic form" generated is perfectly adapted to making the physical reality, a pattern forming inorganic and organic world, understandable.

Today fantastic pictures of galaxies taken from the Hubble telescope are available on the Internet (e.g. fig. 31). One can look into deep space and recognizes the galaxies' distinct and peculiar structures. They are the product of self-organization, which in a time-oriented world is even possible where the initial density of stardust is very low. A much greater variety of patterns and forms is provided by biological structures which take advantage of the highly flexible and rich chemistry of carbon. Also biological evolution is a straightforward process when a fundamental energy- driven time orientation is assumed.

Quite characteristic for time orientation and self-organisation are systems which function between order and chaos. Already towards the end of the 19th century the famous French mathematician Poincaré stumbled over this phenomenon. While Newton's understanding of gravity allows the perfect calculation of a two-body problem, a moon circulating around a planet, the additional perturbation from another planet causes unforeseen complications. Small changes in perturbation can dramatically affect the trajectories of celestial bodies. Under certain conditions a moon can be expulsed from its orbit. This is actually seen within the broad ring of Saturn. In certain ring positions, no moons circulate. There are rings of void. Potential moons here became victims of chaos. Another example of a quite simple, but chaotically reacting mechanical system is a double pendulum, one pendulum attached to another one via the sphere. Its movements end up in chaotic behaviour. What does this mean? Chaotic systems are known to require feedback processes. And feedback processes require the existence of a time orientation. Time invertible basic laws should not yield time orientation, but the pendulum feels the orientation of time without a statistical time arrow. Such an arrow is simply not involved in the pendulum movements. The fact that quite simple systems can develop into time-oriented chaos for me is an indication of fundamental time orientation, as derivable from a time-oriented energy and the principle of least action (chapter 10).

In the meantime, examples of systems existing between order and chaos are known from practically all fields of nature and science. Animal populations can oscillate between order and chaos, as well as a heart subject to physiological problems, a weather phenomenon or a brain between consciousness and sleep. The interplay of order and chaos seems to be fundamental when energy turnover and an oriented time are involved. Interestingly, numerous civilizations of man realized that instinctively early on. They considered chaos as the original substance of the universe, from which some creative act could produce order. It was an order which, when not properly handled, could return into chaos. In Babylonian mythology the chaos was named Tiamat, in Egyptian mythology it was Nut. Christianity remembers the origin before creation as desolate and empty. God or diverse creation personalities then face the chaos and proceeded to act. The ancient Ainu tribe in northern Japan invokes a bird, a wagtail, which patters around on its feet to get order out of chaos.

I mention this relation between chaos and order because energy itself also seems to be subject to such behaviour (fig. 34). It exists in the form of "ordered", free energy, which can perform work and in the form of "chaotic" energy, which corresponds to the entropy product of energy. The latter can no longer be used to perform work. Information can, however, mediate between both. I interpret this interesting coincidence as a further indication for a time-oriented and self-organizing property of energy as I postulated it and as a contradiction to a "sleeping" energy, which is presently applied in physics and which is not related to change, order and chaos. Only distant from equilibrium is such an interplay between "order" and "chaos" to be expected.

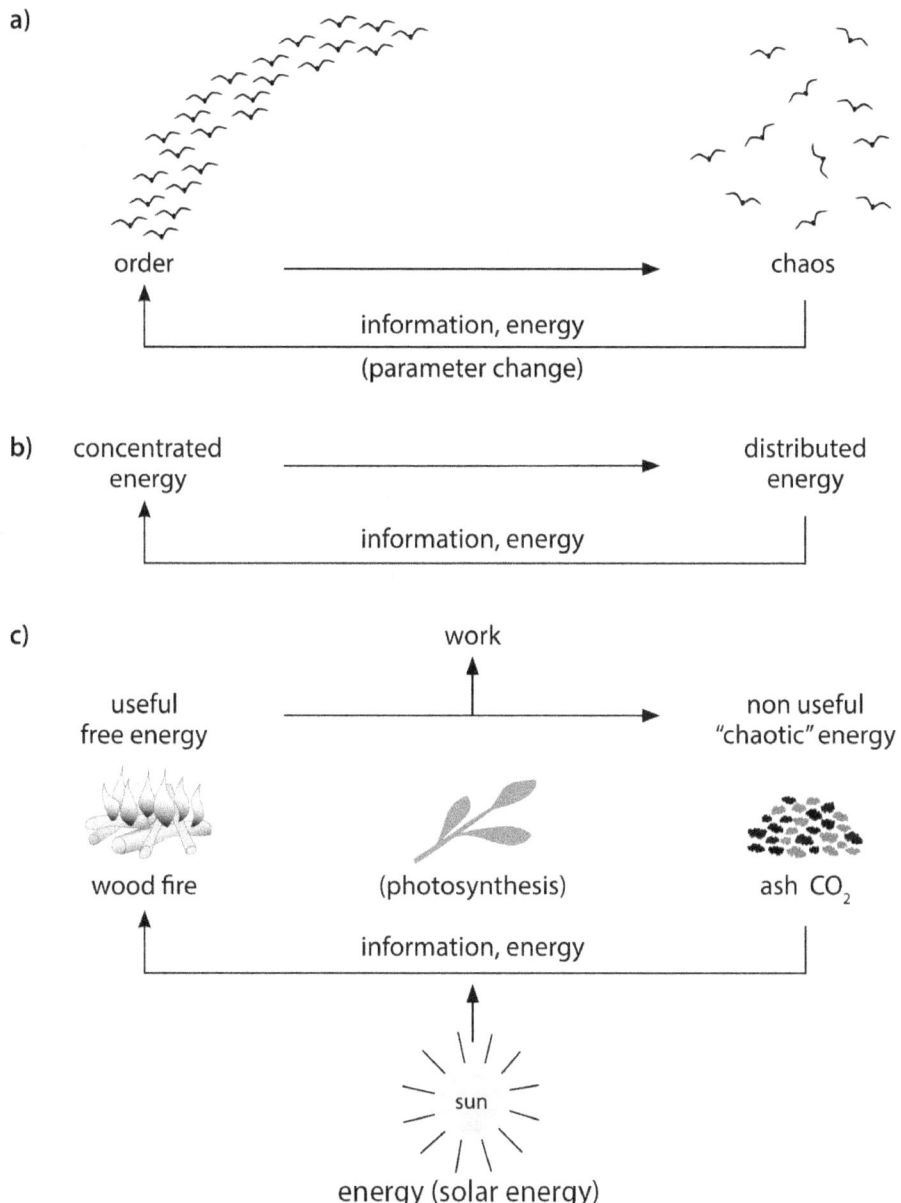

Fig. 34. Self-organized systems can change between order and chaos. Just a modification of parameters (information) is required (a). Information can also mediate between the concentrated energy of a particle and the distributed energy of a wave (claim by this work) (b). In a similar way free (ordered) energy (e.g. wood) can convert into entropic (chaotic) energy (e.g. ash, low temperature heat and carbon dioxide)(c). Information (process of photosynthesis) with an energy input (solar light) can also mediate in this case. This underlines the claim that energy itself is a dynamic and time-oriented phenomenon and shows, in an example, how concentrated energy can be restored from a distributed, chaotic type.

27 Evolution: the aim is high energy turnover and mind

The mechanism of biologic evolution has been studied by generations of scientists. They agreed that evolution has no aim. It is not goal-oriented. Mutations and survival of the fittest are just the mechanisms through which it works and leads to gradual changes and better surviving individuals. Mutations were originally assumed to be arbitrary, but this view gradually changed. Modern evolutionary biologists limit possible genetic variations by considering that the environment has an effect on the probability of mutations.

The basic scientific claim that there is really no aim in evolution is frequently questioned with the argument that there is an obvious development towards greater perfection and better performance. Creationists reject the interpretation of an aimless evolution and argue for acts of divine creation.

Past considerations on evolution had to rely on the world-view of physics: time is a simple ordering parameter and timeflow a purely statistical process. What can the new approach discussed here say about the strategy of evolution?

First, it should be recalled that with dynamic energy a fundamental time arrow is present and self-organization processes in the presence of feedback mechanisms are a straightforward consequence. When environmental conditions permit, and when feedback or auto-catalytic processes develop, local order will build up at the expense of entropy production, the generation of disorder. In a given environment there will be a certain amount of energy available. This will be energy of geochemical or solar origin. Processes occurring in such an environment will compete for this energy. Those that are able to grasp the energy will proceed and generate changes by converting this energy. It is important to note now that self-organized processes have typically the property to convert energy faster and more efficiently than regular not feedback coupled and not auto-catalysed processes. The main reason for this is that reaction products are involved again in the initial reaction thereby accelerating it. In the presence of sufficient energy, self-organized processes will consequently better succeed in harvesting energy than will competing regular processes. Since the turnover of energy generates action, the flux of time, which is visible in the form of changes, will predominantly be caused by such self-organized, living processes. However, different self-organized systems will compete amongst themselves for energy and even more sophisticated self-organised mechanisms will finally result and win the competition, because they will harvest energy faster than others.

We can easily witness the results of this process around us. A modern field of agricultural crops will today harvest a maximum of solar energy and yield a correspondingly high crop output. On barren land this would not have been the case. In nature, self-organizing, living systems did not remain passive and immobile with respect to energy. They started hunting for energy in the form of animals and collecting it from a larger area. Animals were feeding on energy harvesting plants or hunted other animals for energy. New energy fluxes were thus activated. Efficient energy fluxes were important for evolution. Mutations, new ideas, are not very useful if energy is not available to implement them. Energy and mutations in biological systems show more or less the same interplay as money and ideas in economy. New ideas, mutations, require money, or energy, for their realization. Ideas alone without the money for their implementation are often wasted. On the other hand, ideas may more easily be generated in the presence of money, since research and development activity can be activated and paid for. Similarly, abundant energy for survival allowed new mutations to develop.

Therefore, it is not mutations alone, but the combination of mutations with available energy fluxes may be the relevant criterion for efficient evolution. For economy, it would be the combination of money with new ideas.

In a dynamic energy world energy consuming species will increasingly compete for energy. It will be, as I have indicated, their energy conversion activity, which shapes the arrow of time. Changes, which timeflow visualizes, require this energy turnover. Mutations will still be a critical factor in evolution, but all together there will be a faster and faster turnover of energy within the constraints of the systems involved, when increasingly energy-active systems compete. Owing to growing energy turnover, animals have evolved to a huge size as dinosaurs and the currently still living blue whales show, which means they were or are able to devour huge amounts of food. Primitive man, a man living in a natural environment, consumes ten times more energy compared with a meat-hunting crocodile of similar size. This is mostly because a crocodile is a cold-blooded animal and lives with the heat from the environment. Man has gradually learned to get access to richer energy sources and advanced technologies through his mind and intelligence. A modern man in Europe and America now consumes 60 to 120 times more energy than a primitive man. He does it by applying the technical experience of industrial society and its skill in collecting energy from the environment. It can therefore be concluded that a modern man consumes approximately 1000 times more energy than a similarly sized crocodile, a species which appeared very early in evolution.

In a time-oriented world evolution, which allows competition of increasingly energy consuming species, will favour increasing energy consumption provided the constraints of the environment allow that. As higher evolved species tend to be more energy hungry, and more active in harvesting energy, and because energy conversion generates the changes seen in the environment and perceived as timeflow, there is an aim in evolution. It is a gradual increase in energy consumption, provided the constraints of the environment allow this. Constraints are a very important factor. When rabbits were introduced into Australia, they encountered few constraints in the form of enemies and they multiplied dramatically. Their energy turnover had a significant effect on the evolution of other animals and of the landscape in Australia. Other explosively changing ecosystems, for example, grasshoppers devastating huge regions, can also testify to the dynamics of self-organized systems aimed towards increased energy turnover. Also social-political structures can develop in this direction if restraints are reduced. They can develop aggression aimed at accumulating wealth and territories. The history of Genghis Khan`s empire, the Roman or British empires can provide examples. The accumulation of energy, wealth and territories only worked as long as the given restraints permitted it.

It is important to note in this context that we are not speaking here of a time which is just a parameter for measurement. Rather, the changes generated by energy consumption make up the flow of time. This is what is being identified and recognized as evolution in time. Where most of the energy is turned over, there will be the biggest contribution to change. These changes reflect evolution.

A cross-check allows us to test whether our conclusion is correct. Indeed, if our world had remained devoid of biological activity, without forest and light harvesting phytoplankton in the oceans, far fewer photons from the sun would be absorbed and converted, and more reflected back into space or just turned into heat. In reality a significant fraction of solar light is absorbed and utilized by phytoplankton in the sea or by vegetation on land. This energy contributed to evolution and change and is now showing this incredible diversity in nature.

Let us look for more arguments supporting such an unconventional evolution theory.

The first two billion years of the evolution of life seem to have been slow and largely dominated by bacteria and other microorganisms. Around 540 million years ago, during the Cambrian period, an unprecedented explosive period of evolution occurred, during which many existing and many maritime animal species, which later became extinct, evolved. Abundant fossils from that period are found in the Burgess deposits in Canada. The reason for this burst in evolutionary creativity was apparently that a relevant external constraint was removed, the low level of oxygen. The oxygen concentration in the air and in sea water, which served as an electron acceptor for the animal`s energy systems, significantly increased due to photosynthesis. The constraints of the system thus changed. More energy became available. Mutations became much more productive because of access to new energy sources and energy-consuming species flourished. Approximately 67 million years ago meat-eating dinosaurs evolved to the 4-meter-high, 12.4-meter-long and 6.8-ton-heavy Tyrannosaurus Rex. He was the most powerful and successful carnivore. He turned over an enormous amount of energy, mostly by preying on vegetarian dinosaurs. Obviously this dinosaur succeeded by efficiently increasing his energy turnover. During the million years of its existence and of its hunting attacks against peaceful vegetarian dinosaurs, he left a marked fingerprint on time's arrow. It is believed that this species only became extinct because of a meteorite catastrophe.

Then hot blooded animals evolved. Many of those still learned to succeed in forest environments which were extremely low in energy. Examples of these animals are the orang-utang, the sloth or the koala. One species, man, finally started to develop our present civilization. Intelligence helped him to acquire more and more energy. As discussed before, an approximate 1000-fold increase in energy consumption occurred between a similar sized crocodile and a modern man. The latter now harvests energy from most of the plant and animal kingdom and aims at higher and higher energy turnover. When looking at the recent evolution of man, one indeed witnesses a steady increase in energy consumption in spite of energy- saving via improved technologies. Additional energy consumption arises mostly from innovative technologies such as computers, cell phones and infrastructure for housing, transport and communication. In addition, the phenomenon of social networks contributes to increasing energy consumption. The information which is exchanged involves energy turnover and the time applied also has to be paid for with energy. Man has apparently succeeded in loosening constraints, which allows him access to abundant energy, which he also intends to use. The attraction of big cities is basically an attraction provided by a surplus of energy, which facilitates many activities which are not available in small communities. The self-organized systems which man has created provide access to a more intensive energy consumption. Man is apparently aiming towards situations he is already projecting into science fiction movies.

Energy turnover generates changes and changes reflect the flux of time. We experience the timeflow as the turnover of energy, the track of energy turnover. Where energy goes, visible timeflow occurs. What man has activated in the form of energy is finally converted and the track of this energy conversion is experienced as the flow of time or the process of evolution. The evolution of mankind thus also reflects the evolution of continuing and increased energy turnover. And all over the world, over land, sea, in the atmosphere and near-Earth space, the impact of this energy turnover is well visible. Energy conversion generates change and dictates the time, which is experienced as evolution.

Let us now look at evolution from an earlier discussed point of view: the behaviour of self-organized systems (chapter 11), which tend to maximise entropy production. In competing self-organizing systems an increasing energy turnover enables a higher degree of ordering and structuring. Maximum entropy production within the constraints of a system is the result. This means there is a trend towards more complex and more evolved species. When self-organized systems, plants, animals, ecosystems are competing for maximum entropy production within their restraints then their evolution must follow the same tendency. This is actually what we are seeing and what creationists do not expect from an evolution theory without aim, which is controlled only by statistical mutations. Is thus another irrationality eliminated, an evolution towards higher complexity arising from pure statistical mutations?

Everything together fits into a picture of aimed evolution. Evolution is aimed towards an increasingly successful, self-organized harvesting of energy, which, when turned over, drives the observable arrow of time. This energy turnover generates the changes in our environment, which we associate with the flow of time.

What I claim here for evolution is hence exactly what was found out for open, self-organizing systems acting far from equilibrium (chapter 11). They determine evolution by aiming at maximum entropy production. Biological systems and ecosystems, of course, belong to this class of systems and consequently follow its characteristic law, as far as the constraints allow it. Maximum entropy production is the consequence of maximum energy turnover. Since self-organized systems have evolved to dominate the dynamics of the Earth and the universe, their determining thermodynamic law, the drive towards maximum energy turnover or maximum entropy production, shapes the law of evolution.

My claim that evolution has an aim will bring out great controversy, because the science community concerned insists that there is no aim. For this reason it is useful to shortly retrace the arguments given here. A dynamic, time-oriented energy will sooner or later generate self-organized systems such as living species and ecosystems, which will compete for energy. With increasing energy turnover, order and entropy production will increase. The latter will maximise, since entropy production due to energy turnover is rate limiting because of the postulated and derived fundamental law (relation (7)). What is the reason for classical science not recognizing an aim in biological evolution? I believe it was the reliance on time invertible basic phenomena and the understanding of time as a simple ordering parameter. In contrast, evolution pushes away from equilibrium to distant non-equilibrium and consequently has to follow its characteristic laws. The fittest systems, competing with others far from equilibrium, succeed by outperforming the rate of energy turnover or entropy production. And the changes generated carve the trace of time.

Evolution of man is indeed accompanied by marked steps towards increased energy turnover. He has lost body hair, which enabled him to sweat and to chase animals during the hot hours of the day. He thus gained access to an abundant supply of meat. He discovered the advantage of fire for cooking food and for protection. He developed a larger, energy-consuming brain and invested more energy turnover in longer childhood and social structures. Evolution indeed aims at maximum energy turnover and man has succeeded in eliminating earlier constraints and most competition from other living species.

If evolution of life is a straightforward process, why is this not repeated again and again? My interpretation is that evolved life with its increased ability for harvesting energy is too efficient a competitor for energy. Very primitive life can simply not easily survive in the presence of more advanced life. It will be eaten away.

Karl Popper (1979) requires a falsification criterion to be provided for a proposed scientific theory. Evolution is carried on by self-organized systems (human society, animal and plant communities, ecosystems...), which interact and compete with each other. "If it can be shown that self-organized systems do not aim at maximum entropy production within their restraints, then the proposed evolution theory is falsified". Here, of course, the arguments for maximum entropy production from chapter 11 have to be considered.

During a more recent period of two million years of evolution a new powerful phenomenon has been assisting man in increasing the energy turnover for his activities: it is the self-organization of information in his brain. It is the evolution of mind.

28 Self-organization of information: consciousness

Nature allows the development of beings that dispose of a mind. It enables them to recognize their position within their environment and to act in an intelligent way. What I am doing now, searching for a strategy to avoid irrationalities in science concepts, is an example. Consciousness is a phenomenon which cannot be deduced from present science models explaining the evolution of our world. This is a serious problem, because it may indicate that something is insufficient or even wrong with our basic understanding of nature (Nagel, 2014). The inherent possibility of evolving intelligence and consciousness must have been present in the blueprint of nature before man appeared. Therefore, a convincing theory describing nature must include such a possibility in a credible way. The dynamic energy concept discussed here provides it.

The implementation of a fundamental arrow of time into physical phenomena, by postulating and deriving from the principle of least action a "dynamic" energy behaviour, has eliminated all fundamental obstacles towards an understanding of evolution and self-organization of matter. When energy conversion generates action and drives time, there is a "before" and an "after" for feedback processes and this is a critical condition for self-organization of matter and the build-up of order. It is recalled that in conventional physics mathematical manipulation was necessary to explain why entirely time-invertible fundamental processes finally result in a flow of time in one direction only. This manipulation was, as already explained, typically based on the abandonment of information during the calculation process. This should not be done since information has an energy content and omitting energy from a system will profoundly change it. It is for this reason that mathematically simplified processes cannot revert back to the initial conditions. They then appear to be oriented.

Since information has an energy content and as information is proportional to energy, information should, physically and mathematically, also behave in a similar way to energy. Energy is related to matter (via the well known formula $E = mc^2$), and matter can self-organize, as we know from life around us. Starting from a formula describing self-organization of matter, one could therefore, with mathematically clearly defined substitutions and computation steps, derive a formula for the self-organization of information.

Provided the constraints of the system, the boundary conditions, allow it, information should therefore also behave in a similar way to energy and tend to decrease its presence per state. It should be a time-oriented, dynamic quantity. This means that, in the long term, information will again be lost. This is not in contradiction to our everyday experience. Memory is not eternal, and information storage devices will not function forever. As with energy, depending on the constraints of a system, information may be temporarily stored or disseminated into the environment.

As happens with energy and matter, information may consequently also become involved in feedback processes and the build up of self-organization structures, when pushed sufficiently far from equilibrium (fig. 27). For our brain, which handles information via electrochemical processes and a dynamic network of synapses, this means that existing information modes will become engaged in non-linear interactions coupled by feedback processes. This way, locally ordered, or "living" information, information of significantly higher quality, may build up at the expense of a generation of entropy in the form of small and less relevant energy, or information quantities. What meaning has this self-organized information which our "dynamic energy" approach allows? The most straightforward way to find out is to compare it with self-organized matter. It is subject to an analogous, mathematically describable process when involved in self-organization. The feedback processes involved determine the highly non-linear character of the obtained mathematical formalism which reflects sustainment of order, a generation of oscillations, of multi-steady states, or of deterministic chaos. Let us consequently compare a bacterial cell with its well-organized, complex life functions with a liquid, let's say water, which contains only the same chemical elements which are present in the bacterial cell. This may give the impression of the hierarchy between a complex self-organized living system and a not self-organized system containing exactly the same smallest building elements. The difference is merely the missing self-organization process, which, of course, requires a through flow of energy. Since self-organized information may thus reach a hierarchy significantly above that of pure computation with information, and since it may also express strong time orientation, intentionality, there is only one possible explanation for this predicted phenomenon (Tributsch, 2008): it must be the function of consciousness, together with associated, more primitive, but still complex phenomena such as sub-consciousness. I am talking about the mystery of our mind.

Our modern science still has no clear concept of what consciousness and sub-consciousness is. There is not yet a convincing theory, but there are certain criteria, which are given or have to be fulfilled. One is that consciousness is somehow an image of our self, which is able to control and handle it on the basis of skilfully used information from our memory and from the environment. It creates self-awareness and imposes intentions. We know where we are and what we are and what we are doing and intend to do.

A precondition and additional criterion is that sufficient energy has to flow through our information system since a lack of oxygen, which is required for

148

energy conversion in the brain, can lead to a fast breakdown of consciousness. This phenomenon is, for example, seen in high altitude sickness in mountaineering.

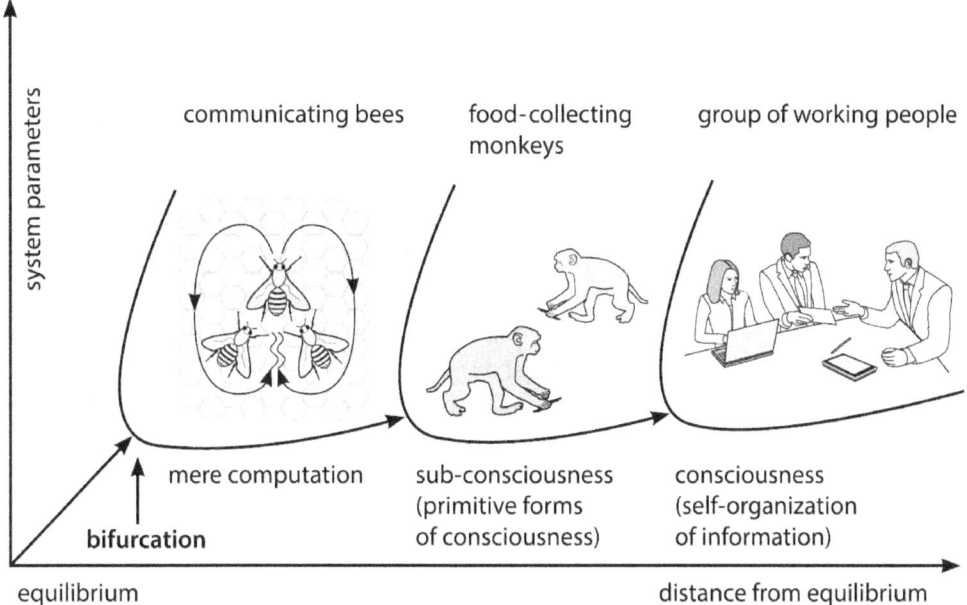

Fig. 35. The diagram shows the influence of the distance from equilibrium for information systems, which is controlled by the through-flow of energy. During evolution more and more energy becomes available to the brain and shifts information activity from mere computation to self-organized computation, which corresponds to more primitive forms of sub-consciousness, and finally consciousness.

The second criterion simply means that the information system has to be pushed reasonably far from equilibrium by allowing a sufficiently large energy flux. This is typical for self-organization processes. They need to be pushed sufficiently far from equilibrium to occur. This also explains why man has evolved a brain which is extremely energy consuming. Consciousness needs a high energy turnover. Approximately 20% of the energy intake through food is consumed by our brain, which accounts for only 2% of our body weight. This is approximately as much as muscles need which account for 40% of our body weight. The brain also consumes 25-40 % of blood sugar.

It is interesting to note here that the extinct dinosaurs, while turning over a lot of energy and existing over a very long time period, did apparently not succeed in energizing their brain towards a simple form of consciousness. Nature had to evolve hot blooded animals, mammals, first, which could regulate their temperature permanently to a sufficiently high value.

An obvious consequence for consciousness is its high energy requirement. A deficiency of energy supply in the brain may contribute to mental problems and, when chronic, to mental illness. Let us take Alzheimer's disease as an example. In this disease there is no competent control and management of computational tasks and the handling of information from the memory and the environment suffers significantly. In this case, the distance from equilibrium in the brain would simply become too small to allow self-organization (fig. 35). Some simple, ordinary computing may still work reasonably well, but a higher hierarchy of computing, self-organized computing and consciousness fails. A characteristic feature of Alzheimer's disease is loss of spatial orientation, even in originally familiar environments. It is interesting to note that simple feedback processes can already lead to elaborate spatial patterns, such as fractal structures (fig. 9). The ability to simulate spatial patterns in the brain may be a precondition for spatial orientation. Such mechanisms may possibly give new insight into such mental illnesses and also explain why they occur when people get older. The energy turnover, the energy availability for the brain is gradually reduced. A reduced metabolic activity in the brain has indeed been observed with Alzheimer patients. The brain system involved operates closer to equilibrium. If one could find a therapeutic way to again increase the flow of energy and the distance from equilibrium (fig. 35), at which the brain works, one could maybe control or, at least, partially handle the problem.

Some general speculations should, in this context, also be expressed on the possible nature of cancer. Life is a self-organized process feeding on energy. If a successful parallel process develops, within the body, due to special environmental or genetic conditions there will be new competition for materials and energy. The competing process, cancer, will either apply a more efficient, different mechanism for energy harvesting, or it will shift the self-organization process further away from equilibrium. This may also accelerate energy and material turnover. The challenge for science should be to understand and control the energy turnover and to decrease the distance from equilibrium. If this consideration is true, there should, however, be an inverse association between cancer (process too far from equilibrium) and Alzheimer disease (process not far enough from equilibrium). It seems that there is evidence for that (Framingham Heart Study, 2012). These are just preliminary, superficial considerations following from the "dynamic energy" approach. They, however, suggest an interesting potential for testing and verification. But let us return to the human brain.

During the last 3 million years the brain size of man has increased, in several phases, threefold from an approximately 450 cubic centimetre chimpanzee brain to a 1300 -1400 cubic centimetre human brain. Man has thereby evolved consciousness. During the same time man has significantly increased his energy turnover. He has left the forests and has adopted a life of meat hunter in the large grass plains using his conscious capacity increasingly for improving his hunting and gathering techniques. Whether the chimpanzee has already started developing consciousness is still not clear. There are, however, scientific reports that claim that chimpanzees show a certain degree of self-awareness. Of course, different levels of consciousness are in principle possible, and human sub-consciousness may be a more basic, simpler and more readily applicable program of self-organization of information. Sub-consciousness is an information-structure like an "automatic mode" of our brain, which controls our situation. It simulates reality and provides an assessment of the environment. It also allows us to perform different activities simultaneously, such as walking, eating and speaking. Sub-consciousness guides our intuition, helps us to recognize faces, and directs emotions such as love and hate. It is tremendously important for many human activities and typically reacts faster than consciousness.

Consciousness as man evolved it is the most sophisticated information program imaginable. It gives a person a personality and active control over his situation. The theory that consciousness controls our computing brain, needs a more profound analysis. The Austrian mathematician Kurt Gödel developed a theorem in 1931, the incompleteness theorem, which became one of the most important statements of modern logic. It can be interpreted to state that a machine controlling another one has to be associated with a more superior hierarchy. The comparison of self-organization of information in relation to information itself with a living bacterium and its chemical components in water gives the adequate answer: a living bacterium functioning via its self-organization structure represents a hierarchy clearly above that of an aqueous solution of chemical substances, which make up the bacterium. Its living activities also clearly demonstrate that it can control chemical activities.

Self-organized information could therefore in principle well control pure information. It is hierarchically definitively placed above pure calculation. It can, therefore, be understood as the mechanism of consciousness. It makes man conscious and helps him to organize his life in a complex environment in a more creative and competent way. It also shapes his intentions and determination.

A statement on the free will of a conscious person can also be made: arbitrarily small information changes, derivable from memory, from unconsciousness or from sensory organs may drastically modify the composition and orientation of the consciously perceived image of our situation. They may induce an entirely new "bifurcation", an entirely new program. The outcome of such processes may be drastically different from person to person and from situation to situation. It could also, however, be synchronized when the external parameters influencing consciousness are intentionally adjusted or manipulated. Free will and its complex properties, when exposed to mass psychology thus seem to be adequately understandable.

As self-organized living species have evolved to and can exist in different stages of complexity, consciousness, as a form of self-organized information, may also have grown and is still growing via evolutionary processes. Psychologists such as C.G. Jung talked of personal sub-consciousness and a collective sub-consciousness, where the heritage of human psychological development is located.

Self-organization of information may have gradually developed parallel to self-organization of matter as soon as information storing and information handling systems, complex nerve networks and brains could be pushed sufficiently far away from equilibrium. When we consider what complex and sophisticated systems self-organization of matter has been creating during evolution we can imagine the potential of self-organization of information. Also reason and intelligence arose from self-organization of information. It is the local order built up into a self-organized information system, which paved the way towards reason and intelligence. In the ecosystem of our planet, man, who has developed consciousness, is also the most intelligent living being. In fact, we need and use this combination of consciousness and intelligence to understand science via the science-related "symbolic forms", which we select. And as we are using these tools which nature gave us, we can hope to understand its mechanisms.

Our collective sub-consciousness and our personal sub-consciousness may just be earlier products of self-organization of information developed within the short 2-3 million years of human evolution. They are still anchored in our brain. The evolution of language may have significantly contributed to the evolution of consciousness. The evolution of consciousness will continue in a similar way as the evolution of matter did towards the large diversity of plant and animal life. Knowing the potential of ordinary evolution, one can easily imagine the potential of highly evolved self-organized information systems in the distant future.

Could, by the way, a computer as we build and use them now, develop consciousness? I believe that this is not possible, since a variable hardware would be required, as our brain has evolved it with its information handling synapses, which are continuously being formed and again disrupted. Such variable "hardware" infrastructure is required for feedback processes. Computers do not yet operate via a variable hardware, but Internet, in principle, does. Here, information links and computational building stones are continuously added, modified and altered. The Internet may develop some form of self-organized information handling. It may generate a hierarchy above that of mere information exchange and develop a kind of consciousness of mankind and a regulatory principle for our planet. Such Internet consciousness will picture an image of our world and dynamically react to changes and modifications to our physical and spiritual environment. Finally, it may tell mankind, of course via continuous human information input, how to behave on the basis of experience and while facing continuously changing environmental and social parameters. Primitive consciousness in the Internet may, for example, help to identify and hunt down manipulated and falsified information, to recognize illegal and criminal activities, and to support initiatives which promise to be favourable for the coexistence and survival of man. Some phenomena in this direction can already be recognized. Political lies and manipulations are repeatedly confronted with counter statements by informed and responsible individuals and groups.

There are other interesting phenomena to be expected from evolution, which concern information. Man with his intelligence is developing better and better information systems. Their technology is facing strong competition. As with the evolution of life, which I claim, is aiming at higher and higher energy turnover (chapter 27), there should also be an evolution towards increasing turnover of information, within the given constraints. What could this mean? I read about a study in Austria, which revealed that school children, on average, check their cell-phone every seven minutes and maintain close contact to numerous friends. We are obviously already experiencing a significant increase of information turnover, which is affecting many aspects of modern life. Most information exchange has peaceful aims. But there is also an aggressive aspect of information turnover, which is on the increase. It has the name "information warfare". It uses information and communication technology to collect tactical information, to spread propaganda and disinformation, for spying, manipulating and disrupting enemy structures or for gaining economic advantages. Also here an increasingly intensive information turnover is observed. There is another remarkable phenomenon: human information turnover and industrial intelligence accelerate energy turnover. Let us just think of the discovery of artificial nitrogen fixation, the Haber - Bosch process one century ago, and its relevance for agriculture and military activities, because fertilizers and explosives could be produced cheaply. Or let us remember the development of transport technology, which has made people much more mobile. Advances in medical research, another example, have led to a significant population increase. They have all contributed to a marked increase in energy turnover by human society, because new or extended activities could be developed. Information and intellectual creativity obviously have a feedback-effect on energy turnover. New self-organized systems arise, which contribute to energy consumption. This again supports the dynamic, time-oriented interpretation of energy and information.

What are the philosophical consequences of the presented concept of consciousness, of superior intelligence and increasing information turnover? These abilities, which we humans enjoy and support, are not mere side phenomena, which have developed by chance, as classical evolution theory would assume. The potential was definitively there before humans existed. And the universe thus has a fundamental tendency to develop a higher life form and an intelligent and conscious spirit. Self-organisation of matter, a straightforward consequence of dynamic energy, would sooner or later have yielded information structures, which in due time would have started to self organise, and to evolve conscious living beings. In localized, favourable locations of the universe life will aim at higher intelligence and at personality-shaping consciousness. Such a conclusion appears to be basically in line with ideas published by Thomas Nagel (Nagel, 2012). He argues that evolution of an intelligent and conscious spirit has to be part of our concept of evolution of nature. The potential to develop such properties should have been there, facilitated by natural laws, before the arrival of man. However Nagel doubts whether a purely naturalistic approach could explain such a phenomenon. On the basis of my restructuring of the energy concept and by giving energy a dynamic property, this was reached here within a naturalistic theory. It basically relies only on physics, chemistry, and biology. And it succeeds in integrating intelligence and fundamental science into a naturalistic foundation. Nagel (2014) argued that a satisfactory theory on consciousness must fulfil two conditions. It should explain (1) why specific organisms have a conscious life, and (2) why conscious life evolved during evolution. Our arguments are clear: dynamic energy and time pushed evolution towards higher energy turnover (1). This enabled the self-organization of information and mind-shaping mechanisms, which opened favourable new opportunities for survival (2).

Since we are talking about human consciousness can we also speculate about a human soul? Science has, in principle, no authority to argue about the soul of a person, but for religions and shamanistic beliefs it is a central matter. Metaphysically inspired personalities have always insisted that something like a soul exists and could perpetuate human existence. Why is such an intuition so strong? Where does science stop and where does religion begin? On the basis of the considerations developed here a person is built up of matter, which is handled and supported by a self-image of information. This should also be the case for information itself and for consciousness, since they have an energy content. This means that an information self-image of a person should exist somehow around the body. It could, theoretically, reassemble a functioning body. We have learned that it can be identified with what we call its gravity. What is its relation to the intuitive concept of a soul? Strictly speaking every object, an animal, a rock or a tree, would have a collective self image. If this could be interpreted as a soul, this would be in line with the Japanese Shinto belief that objects in nature are animated. Also the painted animals in the famous prehistoric caves of Western Europe were animated.

There is another intriguing circumstance for speculation. It has been shown here that biological evolution has an aim: a maximum energy turnover within given constraints. The thermodynamic behaviour of self-organizing systems aims at maximum entropy-production and supports this conclusion. Life and ecosystems are such self-organizing systems. Such properties will automatically favour evolution of consciousness and spirit, since this requires high energy turnover and a large distance from equilibrium. Could it then be that consciousness itself is the real aim of evolution? When I was first confronted with this question, I have to admit that I was shocked. All my feelings towards nature and the destiny of man seemed to be affected. This surprising conclusion contradicts the classical viewpoint of science that higher life is just the result of many coincidences, both on inorganic and organic levels. It also raises another challenging question: why would consciousness and spirit be an aim of evolution? We have again arrived at a border towards religion and philosophy. Man would recognize a sense in attaining a high level of consciousness and spirit, because this would underline his superiority over other domains of nature. He would feel more important and placed in the focus of natural or divine providence. He would also see a sense in the amazing complexity and beauty of nature. A society conscious of such an aim of evolution would be more responsible and would try harder to match imagined expectations. However, an intriguing additional philosophical conclusion is possible - one which would underline the reasonableness of my initiative: if creation of spirit is an aim of evolution, if conscious higher life is expected to understand nature, then the laws governing nature must be rational. There is no room left for irrationality! Irrationality would not support a creative spirit. It would pull it down! It would interfere with its attempt to understand nature and the mystery of life.

The pessimistic philosopher Schopenhauer once commented: "if the world were meaningful, no explanation of its meaning would be required". Evolution of spirit would be a meaningful aim. Man could recognize a task in improving the project creation, in safeguarding an environment where spirit can grow.

How could we name the superior consciousness and intelligence, the extraordinary mind, which is theoretically approachable in the long term in the universe by evolving life? Is it the spirit of nature or is it an image of what religions call God? In his attempt to judge his position in the universe, man has developed surprising imagination. In the Bible one can read: "And God created man in his own image" (1. Moses 1:27). Is the mind developing towards an image of God?

Let us now stop for an instant and ask why my reasoning has conduced us to such unconventional ideas, to entirely new shores. Nobody before seems to have proposed, on a natural scientific basis, that evolution of spirit may be an aim of evolution. Such an aim is not at all deducible from the now preferred model of the universe: It started with a Big Bang explosion, the evolution of life happened as an extreme coincidence, and the end of the universe will be a worn-out, dark nothingness. What was different in the approach implemented here? It is, in fact, much simpler than that adopted by present science with its complicated entanglement in multiple irrationalities on fundamental basis and related to the concept of the universe. The approach presented here essentially relied on the human instinct of causality and rationality given to us by evolution itself. It also relied on simple contemplation of nature. When everything in our environment is changing in one direction, why do we have to stick to historically grown basic scientific laws, which ignore that? This provocative question allowed us to look at the fundamental principle of least action in a different way and revealed its dynamic foundation. Our world is fundamentally irreversible. The rational way nature works also becomes evident when one tries to understand how it applies physics to technology in its biological structures. They are entirely rational in their function. On the basis of such considerations the chance of coming across a relevant concept for the function of evolution, the chance of touching elements of truth, was in our favour. Spirit as an aim of evolution can become a cornerstone of new philosophical considerations. We are a stepping stone closer to the question of what we are and where we are aiming. Man with his intelligence can start thinking about new opportunities and consequences. Living in a universe which has an aim makes a significant difference.

29 The genetic code self-organized

Information in biology is not only handled by neuronal networks, which work via electrochemical processes, but, of course, also by chemical ones, as one knows from the function of DNA (fig. 2), the code of life. Here, alternating and differently combined nucleo-bases, nitrogen containing chemical compounds, make up the strings of the information-handling chemical chain. Each base pair can exist in one of four possible combinations of four bases, known as adenine, thymine, cytosine and guanine. However, when the genome project, developed to understand and handle to code for the benefit of mankind, was in full progress, a shocking insight surfaced. The 2.9 billion base pairs of the human genome only correspond to approximately 750 megabytes of information. This can quite easily be calculated. Each base pair corresponds to two bits of binary data, which can be arranged in four possible states. Together this makes 8 bits, equivalent to one "byte". Accordingly, four base pairs make up one "byte". All together 750 megabytes result. This information capacity is surprisingly small, approximately half the information, which Windows XP by Microsoft needs for its operation. In size, it can also be compared to the file of an ordinary video game. Life is much more complicated than a video game!

It is very difficult to understand how the "blueprint of life", the very complex structural and functional architecture of the human body and how its complex chemistry and neuronal function could be based on such a small amount of stored information. How can the enormous number of body cells and neurons, their supply with nutrients, the immune system, the complex function of the skin, the brain be managed using such a small amount of information totalling 750 megabytes? Such a question is even more justified when it is considered that only 3 per cent of the base pairs were found to be relevant as protein "coding" in genes. A human body contains up to 100 trillion cells, (a one with 14 zeros), 20 billion of them in our brain. Every cell must know to what position of the body it is assigned, which function it should have there and how it should chemically communicate with other cells. Already this is a formidable challenge for information storage and handling. A storage capacity of 750 megabytes simply cannot cope with that.

For a longer time now some researchers have been wondering about the function of heredity and have even been providing unconventional, non-materialistic explanations. A prominent figure is the biologist Rupert Sheldrake, who is proposing a non-physical field of inheritance patterns, named "morphic resonance field" to exist around living beings. It is expected to provide essential additional information, which the body and each cell can tap into in order to function properly (Sheldrake, 1995). It should, for example, provide information on where particular cells have to function within the body.

Such a theory is, of course, not compatible with the present understanding of biological systems. It just creates a support system for biological organisms, which can do everything which we presently do not understand. And I believe, one should not go so far away from established laws with speculations when more realistic approaches are still imaginable.

Here, the just explained considerations on self-organized information can help. A possible materialistic explanation towards largely increased information for heredity is quite easily given. This results from the dynamic energy approach in a straightforward way.

Self-organization of information, as used previously to explain consciousness and the brain, could, of course, also happen with such a chemical information infrastructure as provided by the human genome and the DNA. An immediate indication for that is the finding that DNA sequences show fractal structures and symmetries (Cattani, 2010). It is known that fractal patterns involve feedback processes and self-organization. In fact, Prigogine (Prigogine, Stengers, 1984) already drew attention to the possible role of chemical self-organization for improved information handling on the basis of DNA. Here, not chemical self-organization, but self-organization of information is the focus of our interest.

First, it should not be excluded that the information containing sequences of the double helix, including the mechanisms for reading and implementing the messages, may already have originated from such self-organization of information. Chemical information storing elements may have become involved in feedback-coupled reactions and a build-up of local order in the form of a higher level of chemical information handling. This would explain the great jump in technology of information handling during the early evolution of life. Information simply jumped early into a higher hierarchy of computation. Now, the genome system itself may continue to act as a kind of reference pattern for self-organized reactions, controlling the build-up of "higher ordered" information. It would show up as a tremendously complex higher hierarchy of information branches - bifurcations with information patterns - resulting from the original DNA sequences. They would involve a dramatically increased information storage capacity and the ability to implement and execute the information transforming it into the chemical and structural reality of life. Blocks of similar genes found all over the genome and which create a network of self-similarities may support such mechanisms involving feedback. Ordered information of a higher hierarchy, "living" information, may thus be generated from these feedback-coupled chemical reactions involving the DNA. In fact, this could be a follow-up from the first stepping stone from primitive self-organized matter with simple chemical information handling to advanced information coding and information handling organisms on the basis of DNA. Genes alone as chemical structures, as they are now understood to function, may therefore not be the complete answer towards heredity. However, if they get involved in self-organized information handling in a functioning body, this may be the answer. It is, as previously suggested, like comparing the chemical components, contained in a living cell but dissolved in water with the abilities of the self-organized living cell itself. The difference - a higher hierarchy of organization - is caused by such self-organization. This self-organized information activated with a continuous energy flow from outside, may then build up and control the complicated organism. The information coded in the DNA strings just helps to get the self-organization of information working, which is not in contradiction to practical experience. If one genetically manipulates such a string, one still needs a functioning, energy and infrastructure providing body to create a genetically modified living being. The living body must also provide energy and drive the information structure based on DNA far from equilibrium.

A simple mathematical model on self-organization of information is discussed in Tributsch, (2008). Self-organization in pure chemical reactions is well known and also documented for many biochemical mechanisms in living bodies. An example of an artificial self-organizing chemical reaction is the oscillating and structure-forming Belousov - Zhabotinsky reaction. It shows very impressive oscillating dynamic patterns as long as sufficient chemical energy is present in the solution. It may also convert to chaos or to bi-stable responses.

My claim is, and the dynamic energy approach justifies it, that information modes such as the information coding base pairs of the DNA, get involved in feedback-coupled reactions and yield self-organized information. This occurs far from equilibrium and consumes energy while entropy, in the form of non-available, chaotic energy is generated. But the thus activated, "living" information must be functioning as a higher hierarchy compared with the original DNA information. The new information, implemented in patterns of bifurcations and self-organized dynamic processes, should be different and largely amplified. Both the human genome itself and the surroundings of the living body, which activates numerous additional mechanisms will provide the constellation of parameters, which will control the "living", self-organized chemical information. The "dynamic" energy approach has thus opened new opportunities in understanding biological evolution and the continuity of life.

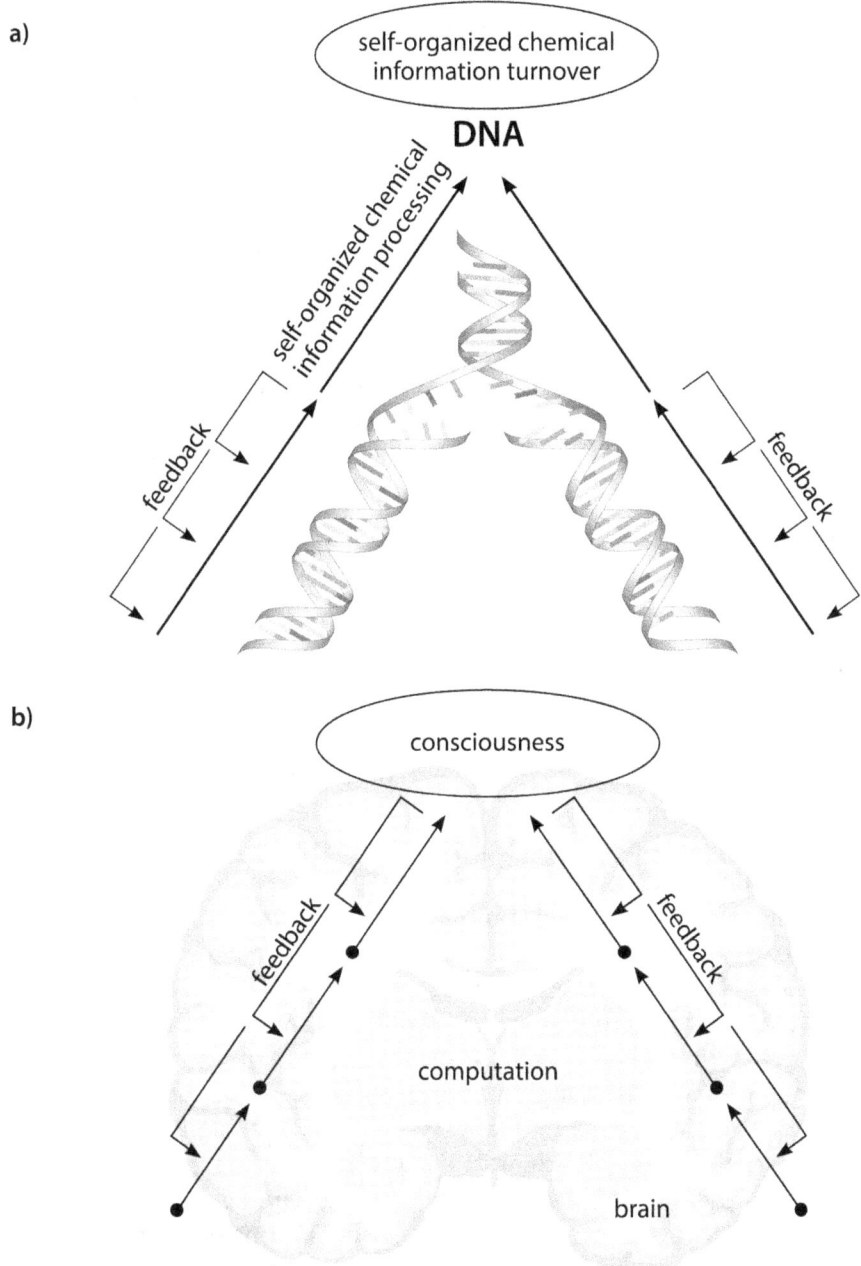

Fig. 36. The difficult step from primitive self-organization to the information storing and transmitting DNA, and to an upgrading of human genome information may have been achieved via self-organization of chemical information (a).
Self-organization of information modes in our brain could similarly have led to consciousness (b).

30 Man and the evolution of the universe

The alternative model or "symbolic form" for eliminating irrationalities from physics proposed here has clear-cut consequences for understanding life and the universe. First, a general theory of relativity, which provides the mathematical basis for space expansion, for inflation of the universe and singularities such as the Big Bang scenario and those of black holes is not needed. It is not empty space, which imposes an always constant light velocity. Constant light velocity has been recognized here to be a local property of photons and not a property of space. Gravity, in addition, is seen as one of the "forces" in nature, which, among other properties, supports energy in its attempt to decrease its presence per state (which is the fundamental postulate for a dynamic energy world). Strictly speaking, however, it is not a force, but information for balancing between the particle-wave duality of matter, with the inbuilt commission to decrease energy per state. It is definitely not a property of space in the presence of matter as the general theory of relativity claims, but present as information around matter engaged in a particle-wave dualism. It is an information self-image of matter.

Within the presently established quantum mechanical understanding, photons have to interact with matter or gravitation to be able to lose energy. In contrast, within the here proposed dynamic energy concept, photons, when spreading out from stellar and galactic objects, can lose free energy on their way and generate entropy in agreement with the second law of thermodynamics. The thus liberated, accumulated entropy is present in the universe in the form of the microwave background radiation. Also self-organization activities of galaxies and of other space structures such as quasars and pulsars, of course, generate entropy. This entropy too will contribute to the microwave background. Its generation at the very low temperature of outer space explains the low energy in form of microwaves. Its origin from spreading starlight and from self-organizing galaxies also explains why there is so much of this low-energy radiation present in space. The reason why we are seeing a redshift in starlight, which increases with the distance of light- emitting stars and galaxies, is this energy loss. It wrongly suggests a very fast spreading universe and an initial Big Bang scenario. The difference in redshift seen with quasars in some galaxies is supposed to arise from differences in energy loss mechanisms. It is not claimed that there is no expansion movement in the universe at all, but all conclusions based on redshift analysis, including speculations on dark energy, should be re-evaluated on the basis of the criticism presented. In order to explain the dramatic redshift in the most distant galaxies, interpreted as near light velocity escape movements, a repulsive dark energy is assumed in standard theory. It is estimated to be so large that it accounts for 70% of the energy of the universe. The dynamic energy concept presented here sees at present no need to discuss dark energy. First the observed strong redshifts should be re-interpreted as an energy loss mechanism of photons (relation (16)). An increasingly faster expansion of the universe may then no longer be evident. Dark energy may be a phantom.

An additional group of existing space theories is based on another doubtful quantum mechanical reasoning. According to the arguments, which I have presented, the uncertainty in quantum mechanics does not reflect a fundamental property of unpredictability and indeterminism, but a deterministic chaotic process exposed during measurement. Energy, popping out from nothing, as is the case in Big Bang scenarios, in inflation theory and in the explanation of microwave background riddles, is therefore, in my opinion, a false and misleading concept. There is no reason to assume that there are effects without causes and a fundamental uncertainty. The uncertainty seen is deterministic, meaning it has a cause. It is based on statistics, arising from chaotic mechanisms.

However, there are also straightforward new opportunities to better understand our universe. The postulate of a dynamic energy that generates a fundamental time arrow has, for example, direct consequences for the evolution of our universe. First, free energy must have somehow been provided long time ago. This energy would have immediately started self-organizing, first generating elementary particles, then atoms and molecules and finally macroscopic structures ranging from geological formations to planets, solar systems, galaxies and aggregates of galaxies. Such a self-organization is not trivial with the present physical understanding, since fundamental laws and processes are defined to be time-invertible and feedback mechanisms cannot develop without a timeflow. Particularly in environments of low gas density as in space, self-organization needs a time orientation, such as that claimed here to be a basic law.

Wherever the environmental parameters are adequate and where water is present, life as we know it will gradually develop. Our earth is a remarkable example. A thin shell of solid soil below a fragile atmosphere, permitting the existence of liquid water between a deadly hot earth interior and a deadly outer space, became sufficiently hospitable for evolving higher life forms. The build-up of order at the expense of entropy generation, via a production of disorder and non-available energy at a maximum rate, will gradually create more and more sophisticated ordered structures and, of course, advanced living organisms. They will not only develop an increasing ability to harvest energy and to build up information structures, but they will be pushed even further away from equilibrium and evolve self-organized information handling. As the living world of plants and animals evolved, a world of self-organized information structures will contribute to diversifying and optimizing evolution. Consciousness and intelligence can significantly contribute to the ability of survival, as we have seen in our world. But it also involves the danger of self-destruction via a destabilization of the environment or via warfare. At present it is difficult to imagine how far such an evolution of matter and mind may proceed. However, it is interesting to note that there will be a twofold evolution pattern in the universe. There will be a gradual accumulation of entropy, non-available, useless and chaotic energy on the one hand, because energy dissipation is needed for self-organization. And there will be an increasingly sophisticated world of organized matter and information on the other hand. Not only organic matter will organize itself, but also inorganic matter, provided external conditions allow it. More and more complex space structures, galaxies and sophisticated stellar objects will form as long as energy is available for sustaining them. This co-production of chaos and order in space is itself a remarkable phenomenon and strongly reminds of the order-chaos mechanisms observed with known self-organization reactions. Only time-oriented processes, as I claim them, can do that.

 Important for new structures and functions is the distance from equilibrium. If the distance from equilibrium can be increased, new so-called "bifurcation" - branches will be reached that may entirely redesign a system (fig. 35). This may also happen when mass increasingly accumulates, turning over more and more energy. A dramatically growing gravitation will then lead to a collapse of matter and to a "black hole". It has been shown that within the "dynamic" energy concept, the function of a black hole can be discussed on an entirely different basis compared to models offered now. As explained before (chapter 23), by joining activity with a quasar, or by developing a quasar, its self-organized energy machine may have bumped into an ultimate way of decreasing energy per state towards maximum entropy production.

There is another intriguing possibility. As already indicated, elementary particles can be seen as products of self-organization, and their wave aspect involves information for a recovery into particles. This information self-image of matter, which has to be located around matter, has been identified with gravitation. Also this kind of information, information on matter, has an energy value and could, in principle, self-organize. Feedback processes would be needed for this, as would be a favourable environment of parameters and energy. We do not know what the complex information infrastructure could be, which would then govern during self-organization. However, in the hierarchy it would clearly stand above normal gravitation and it could act along the same lines but with much higher intensity and it must be supplied by an ongoing dynamic energy flow. Such strong, attractive gravitational forces have apparently been amply localized in the universe. They, for example, act as a kind of glue and keep stars in outer spiral arms of rotating galaxies, which move too fast, from being set free through centrifugal forces. This phenomenon has up to now been attributed to the so-called dark matter, named so because it cannot be seen through telescopes. It is expected to be present in the galaxies in a much higher concentration than visible matter. The latter is estimated to amount to only around 4.5 % of the entire energy - matter budged of the universe. The rest is attributed to dark matter and dark energy. The dark matter cannot yet be identified with detectable elementary particles. The matter of the universe itself is thus estimated to consist up to 85% of this gravity generating dark matter. Self-organized gravitation, or the self-organized form of information, which mediates the particle-wave dualism, is expected to exert a much stronger attraction but would not be ordinary matter. It is still information, a higher hierarchy of information. Do we have a possible   simplified analogy for self-organized information in our technology? It would be like using microwaves and radio waves in the form of laser beams, these waves being the result of self-organization, for telecommunication. They could be used in a much more sophisticated and efficient way, but additional energy for powering the laser technology is needed. Something similar could work with information mediating the particle wave duality of matter, when much information -gravitation- is present. When self-organized, it could act much more powerfully in attracting matter or it could provide additional properties such as structurally organized gravitation.

A new hypothesis is born: dark matter is not matter, but self-organized gravitation. It is the information identified with gravitation, but transformed into a higher hierarchy via feedback processes. Like consciousness in our brain it needs energy to be sustained, but it is still information. Obviously it is for this reason that "dark matter" cannot be seen. It is simply not matter, but it can attract matter because it is gravitation. However, it is a much more effective gravitation, because it has reached a higher hierarchy through self-organization. I recall that this is possible for a straightforward reason: energy and gravitation, identified as information, which depend on energy, are no longer time invertible and can therefore self-organize. As matter can self-organize, gravitation, according to my considerations, can also self organize. I expect that due to energy turnover at the expense of energy dissipation, this resulting super-gravitation becomes much more powerful than ordinary gravitation and can therefore much better stabilize rotating galaxies.

It should be recalled that these considerations entirely contradict Einstein's general relativity theory where a "curved" space-time generates gravitation. Einstein`s gravitation is explained in an entirely different way as a complex property of empty space. Dark matter would here be needed to explain super-gravitation.

If self-organized information of the type discussed here were possible, a kind of simple inorganic intelligence would also be imaginable, concentrated in the centres of galaxies. It would be something beyond information handling and would be able to act with a certain degree of intentionality. Did it help to conduce the evolving mechanisms from energy-devouring black holes to entropy hurling quasars? Such an idea of a primitive inorganic intelligence and a natural creativity may have conduced quite far, but fact is that all our chemical elements, which support our present existence, ultimately originate from nuclear mechanisms and they involve a significant degree of sophistication. Are the elements of the periodic table also products of self-organization of matter and of information in the form of gravity? The cores of atoms with variable ratios of protons and neutrons, stabilized by electron shells, may also be the consequence of self-organization of energy in the form of matter like the formation of elementary particles. Building up order of this kind without self-organization is not evident. As already explained, the quantized orbitals of paired electrons may be the result of self-organized information in the form of gravitation (fig. 20). Only full numbers of wavelength are accommodated in orbitals within a strategy to decrease energy -information- per state. The periodic system of elements, which accommodates atoms in an elaborate order, may indeed be the product of self-organization of energy and matter.

Present theories on the formation of elements see a chance for the formation of elements with higher atomic numbers only in exploding stars (up to atomic number 28, which is nickel) or in very powerful supernova events (higher than atomic number 28) (Nucleo-synthesis, 2014). This would be like explaining evolution of ordered organic life in volcanic explosions. In fact, life is expected to have evolved in aqueous environments. Building up order at the expense of chaotic disorder requires comparatively moderate environments. However, is self-organization known in nuclear reactions? When videos of nuclear bomb explosion tests are examined, sometimes several parallel stripes are seen besides the stem of the mushroom-shaped explosion. One sees patterns of self-organization due to the distance of the system from equilibrium. Inorganic self-organization in nuclear chemical reactions may have contributed to element formation, because the required order for its formation is more readily attainable.

It is a remarkable prospect of this model for the development of our universe that islands will evolve with unsurpassed self-organization of matter and information. The universe is programmed to evolve intelligence and consciousness of a superb quality somewhere where the conditions for energy harvesting are very favourable, while dying in other respects since energy is degrading. What is the philosophical meaning of such a development of the universe? Is this superb intelligence and consciousness already present somewhere in the universe and will human civilization, if it survives for a sufficiently long period, approach it in the long run? During the billion years of existence of our universe, such extremely advanced centres must have developed in abundance. Exceptionally experienced in harvesting energy, they would combine an extremely ordered infrastructure with an extraordinarily advanced knowledge and consciousness. Such a far developed civilization is expected to understand many secrets of our universe.

Currently science does not understand life and human civilization in that way. Life is considered to be a coincidence, tolerated by a favourable constellation of physical laws and environmental parameters.

Science-fiction movies have speculated about such advanced civilisations, which are claimed here to be a highly probable result of evolution. If local conditions allow, and if conditions are sufficiently stable, evolution will proceed in such a direction of maximum entropy production. When energy conversion determines the track of time and when self-organized systems concentrate most energy fluxes on themselves, then evolution will proceed towards increased energy consumption and more and more sophisticated mechanisms of self-organization of matter and information. When confronted with such high-flying expectations for the development of human civilization, it is interesting to recall that different religions including the Christian religion actually nourish the believe that God created man according to his image. Such an instinctive feeling becomes more credible when assuming that man has been put on a track by nature towards greater perfection and that the supposed image of God is approached via a long-term self-organization and evolution of matter and information. When, billions of years into the future, the free energy within space is gradually exhausted and converted into non-useful form, there will still be selected islands with civilizations of high perfection and intelligence. A dying universe will simultaneously be the source of extreme power and intellectual ability. High order will exist close to perfect chaos. Then the remaining civilizations may know the answer to the question how the universe functions.

These considerations brings us back to the proposed inter-conversion between a particle and a wave (relation (11) and fig. 19b). When a particle converts into a distributed wave and thereby also generates chaotic, entropic energy, energy in the form of information has to be set aside to supply a fundamental Maxwell demon. This information will reassemble a particle as a three-dimensional printer could make one. It was a surprising finding that the self-image of matter in the form of information, which would be responsible for that, is what we experience as gravity. Gravity is also fundamental for the dynamics and the fate of the universe. This is also true for all present space theories. It is a fact deduced from astrophysical observations. Now we know that gravitation is an information self-image of matter. Why is this important for the dynamics of the universe? It is a strong indication that exactly the same principle of matter and energy balanced by information is also responsible for the cycles of the universe. An entropy-laden, spread out, old, dark space could be transformed back into a free energy-rich, young universe if sufficient energy in the form of information had been set aside for this purpose at the beginning. The universe should start with a self-image in the form of information and this information could finally reconstruct it. This information, gravitation, is indeed present in space. It could have been set aside from the beginning to guarantee the back transformation of the aging universe. This setting aside of information, which is related to energy, makes the universe formally a self-organized open system with information coming from outside. Physicists may talk of a "galactic demon" in an analogy to the demon operating a gate to allow selectively fast particles to escape from one compartment into another (fig. 10). Alternative names and interpretations may, however, be given to the information power which restores a primordial universe. Primordial energy will gradually convert into distributed energy and entropy. Life and the build-up of galactic structures are, as self-organized mechanisms, coupled to this process. It builds up order locally. Towards the end of this process information, the information self-image of the primordial universe, which was taken aside from primordial energy, will reconvert distributed and chaotic energy,

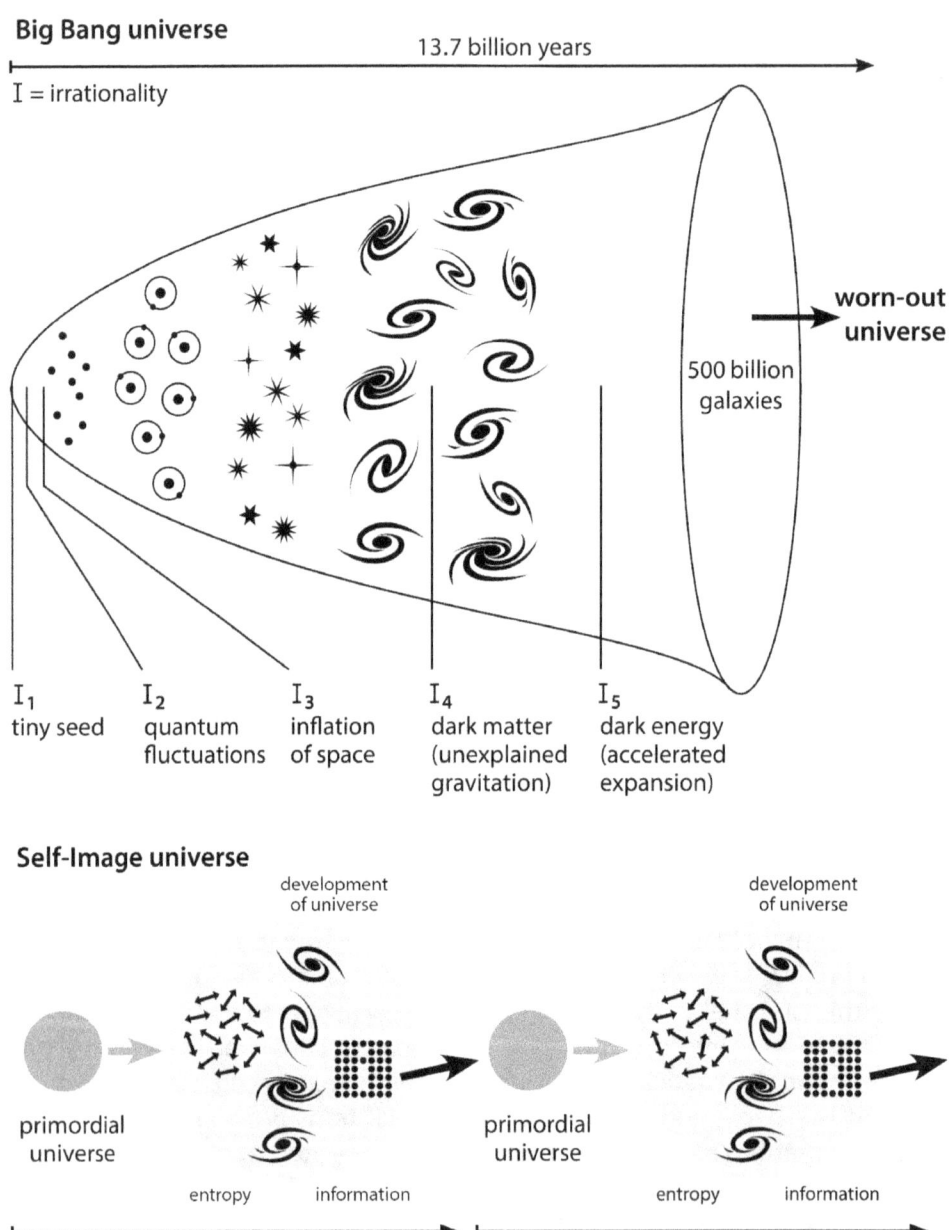

Fig. 37. The Big Bang creation of the universe and its accelerating expansion is the presently established space model. It contains several irrational assumptions (I_1-I_5) (top). The "self-image universe" elaborated upon here, (SI-universe), is rationally understandable. The universe appears to operate in such a way that primordial energy is split up into energy - which is allowed to function and to degrade - and into energy in the form of an information self-image, which will finally restore the primordial situation. If the energy source is sufficiently large, such a process could go on forever repeating itself periodically (bottom).

or entropy, back into free energy-rich primordial energy. This would basically be the same process as proposed for the particle - wave dualism in quantum physics as explained by relation (11) (fig. 19b). The information self-image could be a fundamental property applicable both to the quantum world and to space and could shed light on how the universe functions as a "self-image" universe (SI-universe) (Fig. 37).

If the primordial world contains sufficient energy to provide the information self-image for reassembling it again from the distributed and entropic energy, one cycle could be concluded without a global deterioration. The young free energy-rich universe is allowed to develop while generating entropy. Then energy, which had been reserved before, is used from outside to recover the original situation. This is energetically permitted and avoids the system becoming a "perpetuum mobile" of the second kind. The energy supplied to an imagined mechanism, to a "demon", which is responsible for the back-conversion, prevents that. Such a world-view is not entirely unfamiliar to philosophy. Friedrich Nietzsche, in his late philosophical thinking speculated that the law of energy conservation demands an eternal return of lived reality. With energy in form of information involved this may indeed work, however not in such a detail as Pythagoreans seemed to have assumed two and a half millenia earlier.

On the basis of what has been discussed, the answer given to the question of creation of the universe can indeed be interpreted philosophically. Information to reassemble free energy from entropic energy is needed to reactivate the universe. Information is required as contained in words. We arrive at a puzzling conclusion: for the start of our universe it was more or less as the Bible says: "In the beginning was the word". And the Bible continues: "and the word was with God, and the word was God" (John 1:1). Information was introduced to reactivate chaotic and worn-out entropic energy to start the cycle of the universe we are living in. Information does not have to be conserved, and energy would be conserved through the turned-over information and the energy balance described in relation (11). It has already been applied to eliminate irrationality from quantum physics. The information-driven space cycles could go on forever provided the primordial universe can set aside sufficient energy in the form of information for reassembling the primordial situation. Information, the "word", was present not only at the beginning of the universe, but it also steers the evolution of the universe, which is largely controlled by gravitation. It is information, an information self-image of the primordial world, which activates useful energy at the start of a re-born universe, which then again aims at generating non-useful, chaotic energy in the form of entropy, by allowing energy to decrease its presence per state. This way it is possible to confront the irrationality-laden Big Bang theory for the evolution of the universe with a fractal universe, the SI universe, based on a periodical recreation through an information self-image (fig. 37).

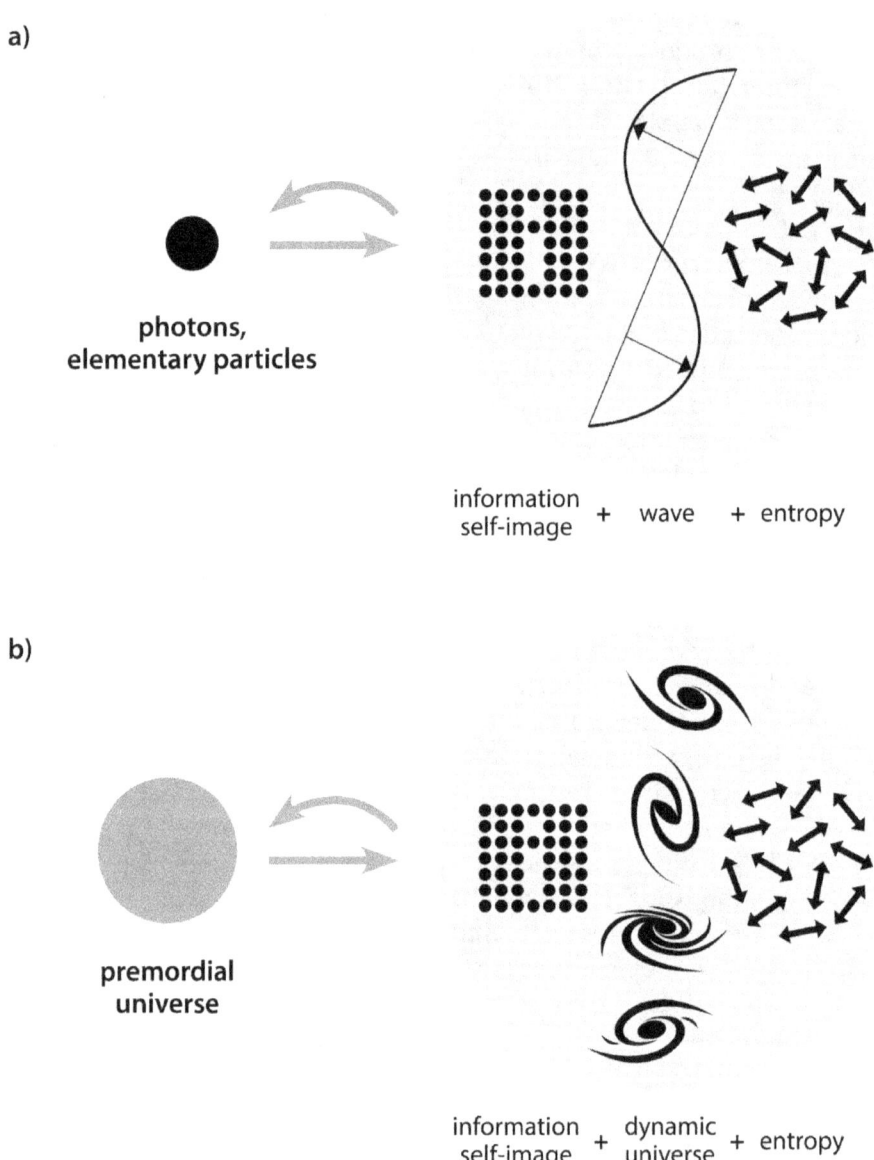

a)

photons,
elementary particles

information + wave + entropy
self-image

b)

premordial
universe

information + dynamic + entropy
self-image universe

Fig. 38. The universe has apparently a fractal, self-similar structure and function: the gigantic space seems to work as a "self-image universe" (b) essentially analogue to the sub-microscopic particle-wave dynamics (a). In both cases information is set aside for the final recovery of the worn-out or spread-out product of energy activity.

31 Mankind's energy challenge: towards success or disaster?

The surprising finding that evolution favours systems and societies which succeed in turning over increasing quantities of energy (chapter 27) has to be evaluated in some detail with respect to the evolution of mankind. This phenomenon, which is entirely in conflict with the present understanding of evolution, arises from the fact that self-organizing systems develop an increased ability to get hold of energy and to generate changes in terms of elements of action. This, however, is recognized as the flow of time as filtered out from action in our brain or our clocks. Systems and organisms, which are able to turn over more energy, get a larger share from the energy available in the environment and contribute more to changes, which make up the timeflow and evolution. In addition, more mutations can be successful when abundant energy is available. The recognized trend towards increasing energy consumption is also supported by the finding (chapter 11) that self-organized systems far from equilibrium will develop towards a maximum entropy production within the restraints given. Maximum entropy production means maximum energy turnover and also the ability to self-organize information and to develop a mind. Human society is such a self-organized system which has evolved a mind.

During biological evolution living species have learned to hunt for energy using improved physiological and neuronal activity. They have also learned to form successful symbiotic associations such as in lichens between algae and fungi. Some species have adopted agriculture technology as, for example, leaf-cutting ants, which grow fungi. Man, of course has evolved consciousness and intelligence and already systematically exploits plants and animals for energy worldwide. It has been emphasized that between an ancient predator such as a crocodile and a modern man there is an approximately 1000 times (600 to 1200 times) difference in energy consumption compared to 10 times in comparison of a crocodile with a primitive man. And modern technologies, such as cell phones, computers and Internet, add to increasing energy consumption. Also the influx of people into big cities is an energy trend. Big cities offer larger energy flows and thus improved opportunities for people.

What man expects from energy is best seen from science fiction movies. The vehicles - which their heroes fly - their tools and weapons, their way of life, all is very energy intensive. And man can well imagine that this will one day occur. As long as organisms and societies generate more successful changes with higher energy turnover, evolution will favour them. And there are consequences. One is the elimination of competition. Due to human civilization and its greed for energy and land, our world is experiencing an unprecedented extermination of species. The other challenge is entropy generation. Energy conversion necessarily leads to entropy generation, or production of disorder. Typically from the products of energy consumption, useful energy is available no longer. Carbon dioxide emission into the air and heat pollution of rivers are characteristic examples for this.

If evolution really has an aim and favours energy-consuming species, then man is facing a serious problem. As early hunters were attracted by lush environments with abundant potential prey, modern man will go on using sophisticated technology to mobilize unexploited energy sources for his profit. While in the past men`s activities were still limited by many natural obstacles and challenges, modern man in a globalized world has largely eliminated competition and is optimizing technology for his profit. He will go on increasing his energy consumption and this may become his destiny. He is risking the destruction of the basis of his existence, which could occur while he is turning over more and more information and becoming increasingly knowledgeable. It is his fate not only to turn over more and more energy, but also more information since information has an energy content. This information tide will change his character, but it is difficult to guess in what direction. It is, however, probable that it will distract him from his main challenges of survival.

The best strategy for the future of man would be to use his spirit and to work towards an increasingly sustainable energy supply and solar energy utilization. The waste energy from sunshine, heat, is again returned to space via radiation so that the environment is not damaged. Such a strategy will give man a breather until advanced technology and increasing discipline will allow him to use abundant energy in an entirely sustainable way. Biological activity of living organisms has shaped the climate of the earth and still stabilizes it. For this reason, man should try to learn from nature in the field of energy technology, taking advantage of energy bionics or bio-mimetics. His aim should be to reintegrate his energy activities into the strategy which nature successfully evolved (Tributsch, 2012, 2008).

To develop a sustainable, well-balanced future, mankind needs a responsible, far-sighted science. In this book, I have criticised science for having partially ended on the wrong track, that of irrationality. With irrational foundations of thinking and acting we cannot adequately cope with the real challenges of our world. These challenges are rationally understandable and should not be veiled in irrational arguments and actions.

An idea on how to deal at least with science-philosophical challenges has been given. However, also the science-sociological challenges have to be addressed. What are the reasons for irrational science models becoming increasingly interesting for the science community and having attracted many young talents? The reasons are obviously the same as those responsible for the recently witnessed financial bubble: new tools and ideas allowed the creation of attractive career opportunities and they favoured professional groups with common interests. In both cases, ordinary people did not really understand what the specialists were doing. Society could learn to deal with such phenomena. It should again try to bring science more down to earth. It could attempt to interest young scientists in the real problems of our survival by giving them adequate career opportunities in relevant areas.

It was speculated previously that not maximum energy turnover, but self-organized information - consciousness, intelligence and mind, which require high energy turnover - may be the actual aim of evolution. Without reflecting on a possible reason for such an aim, it could be argued that exactly such abilities of man would be predestined as strategies against the threat of maximum energy turnover. In other words, man should use his intelligence to cope with environmental problems arising from high energy turnover and all its consequences in different areas. The greenhouse effect, as well as consequences arising from exploiting animals, plants and ecosystems for energy and food production, are challenges for his intellectual abilities. The threat of a fate due to high energy turnover could be reduced and controlled via intellectual creativity. It appears to be the greatest challenge for mankind to activate this creativity and his mind in the direction of survival. Up to now scientific progress, with all its benefits, has not yet demonstrated an ability to guarantee a reasonable long term survival of mankind. Nuclear weapons, toxicity in the environment, overpopulation, exploitation of resources, environmental degradation and climate changes remain significant threats.

32 *Money as an equivalent of energy*

Money cannot be separated from the highlights of human civilization, but nor can it be from its disasters. A few considerations should therefore be spent on this complicated subject. I have mentioned the financial crises and corresponding phenomena. Also some degree of irrationality was involved here. Fictive money and goods, which in reality were actually not available, could make big profit in real money. However, they could also lead a bank into financial disaster. Here also more rationality is needed and there is an interesting link between energy and money: a new theory of energy can also provide new information on the dynamics of money turnover, since money has long been recognized as the equivalent of energy in the economy. Since today nobody really seems to understand the dynamics of money very well, as the ongoing finance turbulences show, this will be an interesting experiment and also a possible test for the reasonableness of the "dynamic" energy approach.

That money is the equivalent of energy in economy can be easily realized. It pays for services, work and materials for which energy has to be activated. And energy has its price. But can we learn more about money if we understand energy differently from science currently? There are basically two factors which are innovative in our energy approach. One is that energy is a dynamic, not passive quantity. It tends to decrease its presence per state and thereby generates action. The other one is that energy can be traded in the form of information which is related to and proportional to energy. It is a kind of self-image of energy. This we have learned when trying to understand the behaviour of energy in quantum physics where it can exist in the form of particle or wave.

Does our money system effectively work, when money "decreases its presence per state"? I would say so, because money on a typical account is exposed to inflation as well as to spending for buying goods and services. The classical concept of energy is that it can do work, but has no interest. The analogy, as described in an example by Einstein, is that of a beggar, who is actually a millionaire, but nobody knows (fig. 12). Such beggars are very rare and if everyone behaved like that showing no interest in spending money, the money system would simply not work. The money system would also not work in the absence of some inflation. Industrial innovations and improvements in productivity have a price and industries have to increase their income to pay for that and to adjust salaries. For the benefit of the economy, the spiral must turn and inflation at a moderate rate must go on in order to avoid economic stagnation. Our money system, therefore, only works if it is time-oriented, exactly as we claim it for energy. Everything that happened with financial systems in the past was indeed irreversible, oriented only in one direction. This is not typical for a beggar who in reality is a millionaire. Our money system would be better represented by a person who shows his wealth through his house, his car, his clothing and his general living standard. Our dynamic energy interpretation, the understanding of an energy, that has not only the potential, but an interest in generating changes, indeed gives a more characteristic description of activities around money.

In fact, only money, which is spent, secures the value of money and a healthy economy. It only flourishes when private persons and companies are interested in spending money. A modest inflation, a situation where the value of money gradually decreases, stimulates the interest in spending money. All these observations underline the conclusion that a dynamic energy, an energy which tends to decrease its presence per state, is a much better analogy for the role money plays in healthy economies, than a classical "sleeping" energy. The presently used energy concept in physics of a "sleeping" energy is not at all suitable for describing an active economy. It describes a failing economy where money cannot fulfil promises. The dynamic energy concept is the actual equivalent of a functioning money economy. Such an economy is always time-oriented. It develops only in one direction. In chapter 28 it was shown that in biological evolution there is an aim in the direction of increasing energy turnover. Also human societies evolve, and there should also consequently be a tendency towards increased money spending within the limits of the system. This can actually be confirmed when looking at national budgets and debts. It is also clear from experience that spending and debts increase when restraints are relaxed. This is indeed well known from many budget discussions.

Based on such an insight we may have a chance to also learn something about the money economy, which is not based on real values but on speculation.

33 What is the equivalent of stockbroking and financial economy?

An important cause for the present financial crises is apparently widespread speculation. Too much wealth in money is not invested into the real economy, but in apparently more attractive investments, such as hedge-funds, derivatives and currencies, which are part of the financial economy. In 2008 the financial assets of three leading banks in Iceland alone increased to ten times the gross domestic product (GDP) of this country. These financial assets are no longer based on real values, but on expectations, promises and speculations. Worldwide, in 2008 they already amounted to 99.6% of all investments. That is, the volume of speculation in the "financial" economy was approximately 250 times larger than the money volume, based on the "real" economy. They deal with promises to get money, with the hedging of these promises or with bets on currency developments. As long as people trust in such investments, they can apparently yield profits. In 2009, at the Chicago Mercantile Exchange (CME) alone, speculations on wheat amounted to 46 times the value of wheat actually traded. The speculations in corn exceeded 24 times the value of traded corn. Speculation on cotton, for example, already starts a year prior to its harvest. Relevant political or weather-related information is highly appreciated and valued. In such speculative financial activities it is crucial to obtain significant information. However, it is known that well-placed misleading information can also be relevant and may cause drastic changes of financial values. Banking institutions engage mathematicians to calculate maximum benefits from small fluctuations. Powerful computers react to the slightest changes within seconds. Sometimes trading is delayed in expectation of greater profits and manipulated information is used to influence expectations. Recently, traders have tried to manipulate results more and more frequently by communicating their opinion on Twitter. Investors read that and react.

If money is equivalent to energy, what is the equivalent of money speculations which are not based on real money, but on information related to money? What is the equivalent of information on money? How is it, for example, possible to sell financial assets without possessing them as is done in "short sale" activities?

It is remarkable that the dynamic energy theory actually considers information on energy (equivalent to information on money) as part of energy properties. Such information is a self-image of energy or matter and has been used to reconvert distributed energy into concentrated energy in formula (11). From information of this kind, energy properties (real economy) can be recovered and controlled.

It has also been demonstrated above that information can self-organize as matter can and well-organized information structures can result from this. The stability and properties of these structures depend on parameters characterizing them.

Financial activities and speculations outside the real economy are apparently products of such "self-organized information". Our dynamic energy concept has consequently created quite a convincing scheme to describe real and financial economy and for distinguishing between them. As there is energy and information on energy to be considered in physics, in economy it is money and information on money. Do such "symbolic forms" also allow us to understand the consequences? A problem and challenge with such a virtual economy is that those investing in such speculations finally want to benefit from real money. In addition, when there is a crash, taxpayers have to save banks with real money and this real money is much more scarce that fictive money, or information on money. If speculations escalate, difficulties seem to be inevitable. Can we, in terms of energy, give a simple example to distinguish a real economy from a virtual one based on self-organized information. A real energy economy could be represented by a peasant in Egypt, a fellah, who works his land and produces food. An energy economy based on self-organized information in a financial economy could be represented by a pharaoh who intends to build his pyramid. He convinces the peasants via his priests that building the pyramid will please gods and bring fertility to the country. Therefore, they work for the pharaoh building the pyramid and donate their share from the real economy.

By betting on the outcome of the grain harvest, speculating investors finally take money from the income of the peasants that grow the grain. They take advantage of a self-organized information system which allows them to keep track of the harvest and get hold of their share. Additional information, for example on an imminent draught, can increase the financial benefit. Let us choose other examples of a self-organized information structure from our present economy. Cell phones give opportunities to exchange and to access information. By providing the self-organized infrastructure, communication companies can extract money and by making the tools more and more interesting for the young generation, they can increase the profit. Another example in which information can activate real money is social networks. They provide a self-organized information structure which people use. The companies behind earn their income by permitting advertisements. The expenses are, of course, finally paid by consumers that buy their products. If too many fictive money speculations are permitted within the financial economy, they may drain too much real money from the real economy. It would be a good idea to handle and tax them like lottery games. A very problematic type of speculation is that of involving countries with severe political problems. Large investors can see an advantage in leading a country into chaos in order to get hold of its valuable assets and can get involved in political manipulation. Misinformation and provoked incidences can become part of financial initiatives. The population concerned, of course, pays the price. Since these information structures all aim at extracting money from the real economy, they have to be controlled or their volume will have to be kept in a healthy balance. A pharaoh should extract only part of the working ability, part of the energy, of his farmers for his personal, religious and political purposes. Otherwise the economy would collapse.

The notion that information on money and its self-organization patterns can be used to deal with money (energy) and to earn or lose money is an interesting aspect of the energy theory discussed here. It may help to visualize the complex flow patterns of money. From the proposed new interpretation of quantum theory it could be concluded that "reality can only be understood logically when energy (matter) and information about energy (matter) are jointly considered" (relation 11a). In analogy, and turning to the financial economy, it may be concluded that "it can only rationally be understood if money and information on money are considered jointly ". Such a conclusion sounds quite interesting, but it may take some time to reflect more deeply on its meaning.

To learn more about information in relation to financial activities is not the subject of this book. It is a complex subject which requires careful consideration.

34 Irrationalities that could be eliminated from science theories

It should again be emphasized that my campaign against irrationality in science opens with a simple, but in its consequences profound, alternative definition of energy. Energy should not only have the potential to do work. It should also have the tendency and the determination to do so. Energy should have the tendency to decrease and to minimize its presence per state. This is a process we are experiencing everywhere in our changing environment. Heat does that when spreading into the surrounding space. Any light-absorbing material does that. During light absorption, electrons are forced into a higher state. From there energy is redistributed until many states are involved and energy is present in the form of thermal vibrations, heat, and emitted low-energy radiation. Why is our claim then something new? Fact is that at present no basic physical law exists which claims a fundamental time orientation in physical processes and energy has no relation to change. It has the capacity to perform work, but no interest, but our postulate claims that energy is related to time and changes. In fact, it is not a postulate. It was shown here to be derivable from the fundamental principle of "least action". It is recalled that this principle says that the integral, the summing up of the product of energy and time, from the beginning to the end of its path, aims at approaching a minimum (fig. 16). When subdividing this integral into infinitesimally small fractions these also have to minimise. And this can only happen when energy and time, as well as their product, action, are able to change towards a minimum. They have to be dynamic. When dealing with the principle of least action, classical science has not recognized this fundamental property. The reason is obviously that it is using an energy - which is just a sequence of numbers - and a time - which is just an ordering parameter. Science could not understand the real dynamic meaning of the principle of least action, even though it has become so important for deriving fundamental laws.

A dynamic, oriented energy and an oriented time are therefore a paradigm change in physics, and the energy interpretation presented here leads to the claim that it implements the principle of least action in a correct way. It is even claimed that the principle of least action is the equivalent of our dynamic energy statement. This has a chain of consequences.

The first is that the second law of thermodynamics immediately follows. It states that entropy, disorder in the form of non-usable energy increases in a closed volume. It stops, however, to be a purely empirical postulate. It can now be derived from a more basic postulate, the new symbolic form. This was not possible up to now. Also a law for irreversible thermodynamics follows. In the presence of time orientation, feedback processes are allowed and self-organization is a straightforward consequence. The classical challenge of deriving irreversible processes from totally time-invertible fundamental mechanisms through mathematical manipulation which abandons information has lost its relevance. Self-organization of matter has thus become a straightforward process. In the non-linear range of irreversible thermodynamics where this occurs, maximum entropy production is expected.

The dynamic energy paradigm applied to quantum physics first examines the particle-wave duality. It understands it in such a way that the particle tries to decrease its presence per state by expanding into a wave. Energy is not turned over for work, and to sustain the particle-wave dualism information is required to reconvert the wave again into a particle. This inter-conversion duality between particle and wave is the new quantum state. It should be emphasized here that in handling the duality of particle and wave, classical quantum theory has made a mistake. It overlooked the relevance of space for energy. Concentrated and diluted energy are not identical. Here, with this mistake, which overlooks the fact that expansion of energy generates entropy, irrationality in quantum theory starts. In fact, information must mediate the inter-conversion between particle and wave and one perfectly understands now, via the information involved why the second slit is recognized in the two-slit experiment. It is no longer necessary to assume that the particle is simultaneously at two locations when passing the double-slit screen. It is the information involved in this dynamic state, which allows us to understand the role of the second slit. The non-locality paradox is eliminated. The wave is no longer related to the probability of localizing the presence of the particle. It is merely the result of the particle's tendency to decrease its presence per state by distributing the energy. In the same way, quantum correlation - the strange property of quantum particles to communicate between each other over larger distances - can be understood rationally. When a particle splits up, information also has to be split up and to be restructured. This is apparently not an easy job when natural laws like the conservation of total angular momentum have to be respected. Therefore, a   contribution of joint information still maintains a link between separating quantum-correlated particles (fig. 28).

Dynamic self-organization, possible because of the time-orientation of energetic processes, is also used to explain the manifold of elementary particles and their behaviour. Subject to a fundamental time-orientation and in the presence of feedback processes, energy can self-organize and yields particles of varied stability and properties. In principle they could be understood like virions, non active viruses, of different sizes and properties. When elementary particles or atoms decay, this may occur with statistical probability. But the process is not arbitrary, it is deterministic. It arises from deterministic chaos. The non-causality paradox of quantum physics is thus eliminated.

The continuous reconstruction of a light particle from the wave through the information self-image finally opens a possible way to explain the constancy of light velocity independent of the velocity of the reference frame. Photons could be seen as programmed to reassemble from information always at constant light velocity. The mechanism is analogous to that which controls a three-dimensional printer converting information into a functional product, for example, a car which can drive at a given speed. No special four-dimensional space property is needed to force light into an always constant light velocity which is experimentally observed. The constant light velocity paradox would thus be eliminated. Independent of velocities, information can always provide a comparable function, in this case constant light velocity. The new interpretation of constant light velocity can be understood intuitively.

Such a local, photon-centered cause for the constant light velocity has quite dramatic consequences. There is no need to assume that space is responsible for constant light velocity. Consequently, there is no need to claim a four-dimensional space-time. The space-time paradox is thus eliminated, a paradox which manipulates the size of objects and the time it feels. The special and general theories of relativity have then just limited significance as mathematical tools for describing quantum phenomena, subject to measurements dependent on light velocity. They describe time and length dilation for quantum phenomena where energy is not converted and energy is not related to time.

When, however, energy is converted and energy drives time, both have to be transformed as action and action is invariant in the theory of relativity. Energy converting systems can therefore not travel in time. The paradox of time travel is thus eliminated.

Currently in physics, time does not pass for a photon, which travels at light velocity. It can only lose energy when interacting with matter or gravity. This is different for the new dynamic energy postulate. It not only allows but requires dissipation of energy. Light spreading from stars and galaxies into space must produce entropic, non-usable energy, because it is an irreversible process which generates entropy. Quantum physics again overlooked the effect of space on energy and does not allow spreading photons to generate entropy. This deficiency is corrected here. The phenomenon of energy loss and entropy generation, which is in fact required by the second law of thermodynamics, is expected to produce the redshift of light which increases with the distance from the light sources. The redshift of starlight, which is now considered to be a consequence of the Big Bang expansion of the universe, is merely caused by the energy loss of spreading photons. This explains why we seem to be at the centre of quite a homogeneous universe. We just see the energy loss of light from space objects at varying distances. If this is confirmed, the structure and dynamics of the universe have to be re-evaluated.

Periodical patterns in the redshift of starlight should indicate periodic energy dissipation. The background microwave radiation is considered to be the entropic energy accumulated in space. Large black holes in combination with quasars are understood to be self-organized structures aiming at maximum entropy production. Four irrationalities are eliminated or at least questioned: the Big Bang origin of our vast universe, the inflation of space postulated to explain its homogeneity, dark energy and dark matter. The redshift phenomenon observed has to be reinterpreted accordingly.

In fact, the main paradoxes and irrationalities in physical science could be avoided simply by changing the energy paradigm, by insisting on the role of space for energy, and by considering a self-image of information, which is a consequence of this. Fig. 37 shows how the new "symbolic form" for physical science, based on these three new basic foundations, works,. It also indicates where paradoxes are eliminated, and visualizes how new explanations are related to the fundamental new postulates.

This way the symbolic form for understanding physical reality was redesigned. There are also quite significant consequences of the new approach discussed in other fields of science. In contrast to the present view, evolution has an aim. Increasingly energy-consuming self-organized systems compete for energy and accelerate energy turnover. This coincides with the behaviour of self-organized systems far from equilibrium. They aim at maximum entropy production, which takes place when maximum energy is turned over. Since this energy turnover causes the changes we recognise as timeflow, evolution also seems to be accelerating. Time itself does not proceed faster, because it is calibrated, but more changes happen in the time when more energy is turned over which is also stimulating new inventions. Such an apparently accelerating evolution is easily recognized when comparing the present time period of human civilization with earlier ones. The same is also true for biological evolution. Two billion years passed with only bacterial life, five hundred million for the evolution of plants and higher animals. It took only two million years for the evolution of man and intelligence, and intelligence and intentionality are characteristic for a dynamic nature. It was shown that such functions can work via a self-organization of information. This mechanism may open new dimensions for understanding human personality and consciousness. In addition, it also has a fundamental meaning for philosophy.

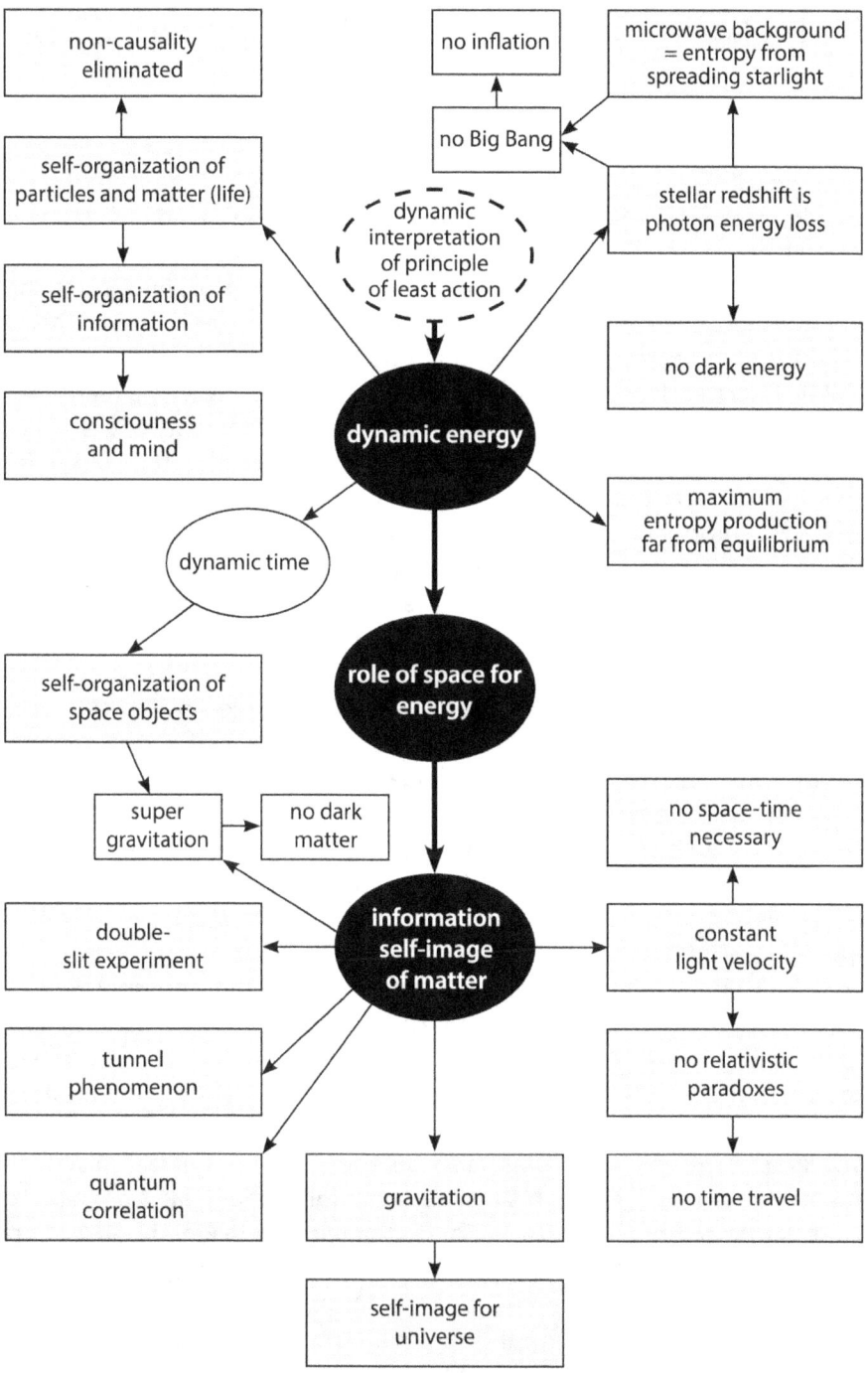

Fig. 39. The new symbolic form is a fundamentally dynamic energy-time world, respecting the role of space for energy and considering an information self-image of matter (three basic claims on black background in diagram). It is shown how from these derived three basic postulates essential conclusions could be deduced.

35 Evolution including mind

One of the significant accomplishments of the "dynamic energy" approach in understanding nature is that it can close the gap between a largely materialistic concept of evolution in nature and the evolution of consciousness and intelligence. Nature allows the development of living beings which have a mind and which attempt to understand nature. They actually succeed in obtaining a reasonable concept of their position in their world. A most important challenge is the question how the mind, consciousness and intelligence are inherent in the universe. One obviously deals with a fundamental aspect of nature. The claim of philosophers such as Nagel (Nagel, 2014) who believes that already before man appeared on our planet, nature must have had the capability of creating his mind, is relevant. And the dynamic energy approach enables that link without any additional assumptions and efforts. As matter (energy) which created life can self-organize, information, which is also based on energy turnover, can do the same. The precondition is only that some degree of order in information structures has to evolve before. The information storage elements involved in self-organization could be chemical, as implemented in the genetic code, or electrochemical, as function in our brain. The result would be a much higher hierarchy of computation. An equilibrium science, time-invertible laws and a time-neutral energy phenomenon could not sustain such a process. The implementation with the dynamic energy concept of a fundamental energy-time drive is decisive for the shift of systems away from equilibrium and for the necessary feedback processes. It is claimed that the model of evolution presented here is much better prepared to describe reality than traditional concepts, not including irrationalities which were inherent to them. The discussed model is, of course, only a sketch which in the future would have to be much better elaborated upon.

An interesting additional subject for further thinking and speculations is also gravitation, which has been identified here with the self-image of information related to the particle-wave duality of matter. In physics it is still a mystery and has been discussed as a very strange force. But in this study gravitation is found to be information, implementing a decrease of free energy. As already mentioned this kind of information should also be able to self-organize, provided enough energy is turned over to push the gravitation system far from equilibrium. A significantly higher and more elaborated gravitation, a kind of "living" gravitation may be the consequence. And gravitation should also be spatially structured like living organisms. This phenomenon may, for example, take place in the centre of galaxies where high concentrations of matter are present and interact. But it may also occur widely distributed over the space, structuring it. Astrophysicists know a phenomenon called "great attractor". It attracts galaxies, including our own. This may be the result of self-organization of gravitation. The resulting super-gravitation will have much more sophisticated properties with respect to the interaction of matter. Above all it is to be expected that the capacity to decrease the energy per state, which this information - gravitation - should implement, will be much more effective. There should be a much higher attraction between matter, but it is perfectly understandable why light can penetrate a super-gravitation zone, though with deviations due to the mirage effect, which is actually observed. This happens similar as an ordinary mirage on earth develops under a special weather condition, which itself is a self-organized phenomenon.

Obviously this is an alternative theory to dark matter since it would explain a high and spatially structured gravitation, which is not balanced by sufficient visible matter. But no additional matter would be needed in this case. There would not be a need to search for invisible new particles, but there must be a source for energy turnover to push the gravitation system far from equilibrium. And if it is a hitherto unrecognized property of gravitation, a self-organization of gravitation, then additional speculations seem possible. Since gravitation was identified as information, such super-gravitation areas, mostly in the centres of galaxies or within clusters of galaxies, should also have abilities towards a higher hierarchy of information handling and a kind of primitive cognition. The different structures of galaxy clusters exhibiting in their centres one, two, a linear chain or aggregated large galaxies, may evidence such abilities of building up order. Are galaxies and galaxy clusters gigantic lifelike inorganic creatures? There is something mysterious out there in space, a kind of inorganic intelligence, which is much more sophisticated than can be expected from a simple Big Bang origin and an increasingly fast expansion of the universe! The complex network of galaxies, observed in dense regions of the universe resembling organic structures, may support such a claim of a superposition of self-organization of gravitation (information) and matter in space (fig. 40).

Gravitation, the information self-image of matter, also plays a central role for the general dynamics of the universe and self-organized gravitation suggests even more sophisticated ways of interaction. The latter not only seems to produce the super-gravitation effects in galaxies, but may also be needed to transform a worn-out universe back to a primordial state in a similar way as a spread out wave is reconverted into a particle. There is mind not only in our brains and in the information structures of unimaginable many living species out in the stellar systems across space, but there is some kind of living mind also out in the hostile and life threatening structures of space itself (fig. 40). To understand its function will remain a long-term challenge for mankind. But such an effort will, hopefully, make man more modest before the miracles present in our universe. Understanding energy and energy-controlled information as a dynamic phenomenon in time would be a first step.

When evolution of the universe aims at maximizing energy turnover in its self-organized systems, in complex space objects and in organic life, then intelligence and consciousness will be a straightforward consequence. This allows us to ask a challenging philosophical question: is it thinkable that the real aim of evolution is then the ultimate evolution of mind? Mind and intelligence have the capacity to

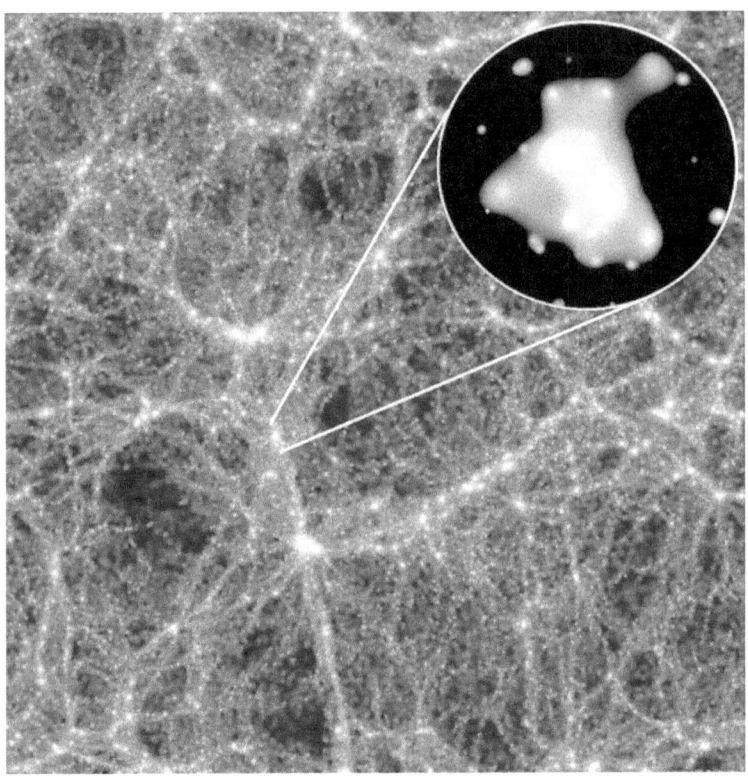

Fig. 40. The galaxies in the universe indeed adopt a complex order of clusters, lines and voids which suggests a fundamental mechanism of self-organization operating over large distances. Time-invertible physics cannot explain this. (An XMM-Newton X-ray image of a galaxy cluster superposed on the network of galaxies in a dense part of the universe) Credit: NASA, Virgo Consortium: Jenkins et al, 1998, http://chandra.harvard.edu/press/05_releases/press_040805.html

enjoy and understand nature in its beauty, but also to control and use it. It is, of course, also an important mechanism towards a diversification of evolution. The vision of such an aimed strategy of the developing universe is a fascinating contrast to the present view in science of higher life as a chance artefact. It supposedly took advantage of a very favourable constellation of essential, but basically accidentally present inorganic and organic conditions for survival. The magnitude of the gravitation constant is an example of an inorganic parameter - the average temperature and the presence of liquid water on the earth surface would be additional ones. Favourable conditions are, of course, necessary but the situation is more complicated in a dynamic energy world. Self-organized systems would gradually evolve towards higher order by increasing the energy turnover and adjusting their function with respect to given external parameters. Let us select examples: when the rotating earth was forming from matter circulating around the sun, its core was melting and stimulating liquid metal flows to self-organise thereby creating the earth magnetic field. This became a very important protective shield for life on earth. Then the atmosphere and climate developed, and oceans formed with a self-organized current pattern, both adapting to developing continents. Embedded in these self-organized systems life developed as an additional self-organized process and finally so did mind. And life, via photosynthesis, had, of course, a strong feedback on the evolution of the climate. Not favourable physical parameters alone, but much more conditions, which self-organization shaped, were spearheading evolution. And they were aimed at maximum entropy production, within the constraints of the systems. This kind of evolution, controlled by laws of self-organization, is proceeding not only on life-supporting planets, but also in hostile space.

The evolution of mind became possible only after other self-organized systems prepared the required environmental conditions. For life it finally became an important tool towards enhanced creativity and diversification. Man is, however, simultaneously faced with the dangerous product of increasing energy conversion. His mind should be the creative tool against the devastating accumulation of entropy from energy turnover in the environment. Mankind is invited to use spirit and intelligence to deal with the problems.

Such circumstances have arrived now on our planet. The chances facilitated by high energy turnover have to be balanced by intelligent and sustainable technologies. This could be a criterion for the survival of advanced civilizations. Only the combination of high energy turnover and a developed mind promises an ongoing sustainable future. Evolution to a high level of energy consumption without a reasonable spirit would lead us into a dead end road. Living organisms or human societies would simply destroy their environment and allow toxic products to accumulate, but equipped with a reasonable spirit they can master their problems and continue evolution. Seen this way evolution of mind is also a tool of evolution. A high standard of evolution can only be reached with a well-developed and constructive mind. There is consequently a challenge for an advanced society to safeguard health and evolution of mind.

A RATIONALLY UNDERSTANDABLE NATURE

36 What could be learned from a dynamic energy world?

This investigation is an enquiry into irrationality in existing models and theories, relating to physical processes which proceed in our world and in space. Are laws of nature themselves irrational, or is it our current knowledge of science, which, not sufficiently understanding nature, is inventing irrationalities and using them to construct theories? One also has to ask: why can mind, consciousness and the intentionality of human activities not yet be derived from basic physical-biological mechanisms? Is it the irrationality of thought models which is blocking the access to understanding? Can it be, for example, that the concept of effects without causes, or the absence of a fundamental time orientation, is preventing us from understanding consciousness? This book shows that a rational approach to the fundamental problems of present physics makes sense. It leads to a sufficiently consistent understanding of nature and to reasonable secondary questions and challenges.

At the beginning of this book some thoughts on rationality in relation to irrationality, as well as on mechanisms for recognising our physical world, were revealed. They should serve to initiate a search for the origins of irrationality. It was hypothesized that information is missing in quantum theory and this has become a source of irrationality. Since information has an energy content, the attention consequently focussed on the concept of energy. It could be incomplete and partially wrong.

A kind of strategic guideline for these studies therefore became critical considerations of our evolved concept on energy. What is missing or wrong and what is the difference in energy concepts applied to classical physics and quantum physics? More than two thousand years ago the development of the concept of energy started as the search for something which remains conserved within all changes occurring in our environment. It turned out to be energy, which during all changes occurring, is conserved. However, a puzzling fact is that our present energy concept has abandoned and lost any relation to change. Energy has the capacity, but no interest in doing work. In physics it has become a neutral quantity without any orientation. Such a peculiar, modern definition of energy was recognized to be a main reason for irrationalities in physical theories. I redefined the energy as dynamic, an energy with the tendency and interest to minimize its presence per state. For me this interpretation of energy is entirely reasonable and also fully compatible with what people have always felt about energy. Someone who has energy is also willing to apply it and to induce changes in his surroundings. I could also quite easily demonstrate that the dynamic energy idea can be derived from a very basic principle. It is the principle of least action, it is a very fundamental principle in nature and is the origin of numerous physical laws which were derived from it. While nobody could explain why nature has adopted this principle of least action, it has become an anchor for understanding many of natures laws. Mathematically, the principle of least action is described by a time-integral over the energy of a system, which has to become a minimum (fig. 15). Classical science relies on an energy which is not linked to time and changes, and has fundamental laws which are time invertible. This science looked at the outcome of the integral, but not at a possible additional fundamental meaning of the principle of least action. The fundamental meaning can be recognized, when the time integral over energy is split up into a large sum of infinitesimally narrow time integrals over energy. They all have to become minimal. What does this mean for energy? The narrow time integrals over energy can become minimal only if the energy minimizes its presence per state, and when an oriented, changeable time is coupled to this behaviour. This is exactly the "dynamic" energy paradigm proposed in this study. It can therefore be claimed that it is derivable from the principle of least action. And, when mathematically performing the minimization of action by differentiation of the least action integral, and searching for the minimum, one indeed arrives at the proposed dynamic energy statement. The principle of least action and the dynamic energy claim are identical: the principle of least action is a dynamic principle and has its origin in an energy which drives time. For the first time it can be explained why nature is so persistently adhering to the principle of least action. It is simply implementing dynamic energy properties. It is telling us that the reality in our environment is on-going changes and not stagnation. As a consequence, the presently applied concept in physics of a "sleeping" energy, an energy not directly related to change, is an oversimplification and has to be abandoned. Such a conclusion is supported by an additional observation in relation to energy. The relation between free, available energy and non-available, entropic energy is like the relation between order and chaos in self-organized systems. Information, such as obtained from a Maxwell demon, with its energy content can mediate between them (fig. 34). Only systems which allow energy to flow through, in a time-oriented environment, are known to develop such properties.

To come back to Einstein's comparison of our classical concept of energy with a beggar who in reality is a millionaire (fig. 12): I convinced the beggar to live like an ordinary rich person and to spend his wealth on every day necessities and the seductions of life as well as on charitable purposes. This eliminated, of course, the mysteries (irrationalities) around the beggar - millionaire. By making his secret public and by exposing his wealth the mysteries around him were clarified. His money became the motor for many activities, and his life became more transparent. The same is expected to happen when "sleeping" energy is changed into "dynamic" energy. Irrationalities disappear and new opportunities for understanding are opened. For example, the second law of thermodynamics which states that in a closed system entropy will maximise, can directly be derived from it. This was, in fact, not possible until now and was a shocking fact for classical physics. Also maximum entropy production can be recognized as a fundamental property of self-organized systems, operating far from equilibrium, such as life, ecosystems or hurricanes. Because they dominate our environment their entropy maximizing behaviour was a first hint at an aim in evolution of life.

Energy has to be understood as a dynamic quantity, which generates changes and drives time via conversion of the energy into a "chaotic", no longer useful form. This energy product contributes to entropy. It is like money which has been spent. It is still there in the surroundings, but not available and useful any more for the spender.

After specifying the applied "dynamic" energy paradigm, which for the first time introduces a fundamental time arrow into physics, the enquiry on energy then continued with a study of irrationalities and paradoxes in quantum processes. First, the adopted fundamental dynamic energy principle had to be applied to quantum physics. Energy should attempt to decrease its presence per state, but quantum states, as long as they function, do not generally allow a system to convert energy. How should, for example, an elementary particle such as an electron or a photon, decrease its content of free energy without interacting with matter or gravitation? Energy could simply spread out into space. Here the idea was born that this is exactly the mysterious twin state of a particle, with its property to appear as a wave. Energy attempts to convert from a particle to a wave. Matter cannot only be described as particle but it is describable also as wave. However in the wave, where the energy is distributed in space, energy has an inferior value and has therefore to be understood differently. Consequently, there is no equivalence of particle and wave, as stated in quantum physics. In contrast, it has to be assumed that dynamic energy converts a particle into the form of a wave. Energy conversion and entropy production is, however, also suppressed by the wave.

The transformation of a particle into a wave and the existence of energy in the form of both a particle and a wave, remains the expression and characteristic property of a quantum state. Re-conversion of the spread wave into the concentrated particle requires information. The quantum state has consequently to be defined in a different way (relation (11)). It includes both, reality in form of particle and wave and information on this reality. What are the consequences and where are irrationalities in traditional quantum physics now coming from? I retrace the strategy I have adopted to identify the origin of irrationalities:

In the form of a hypothesis, quantum paradoxes were interpreted as indications of a lack of information from quantum theory for the observer. Since information has an energy content, the energy concept of quantum physics was consequently seen in contrast to that of classical mechanics, which does not yield paradoxes.

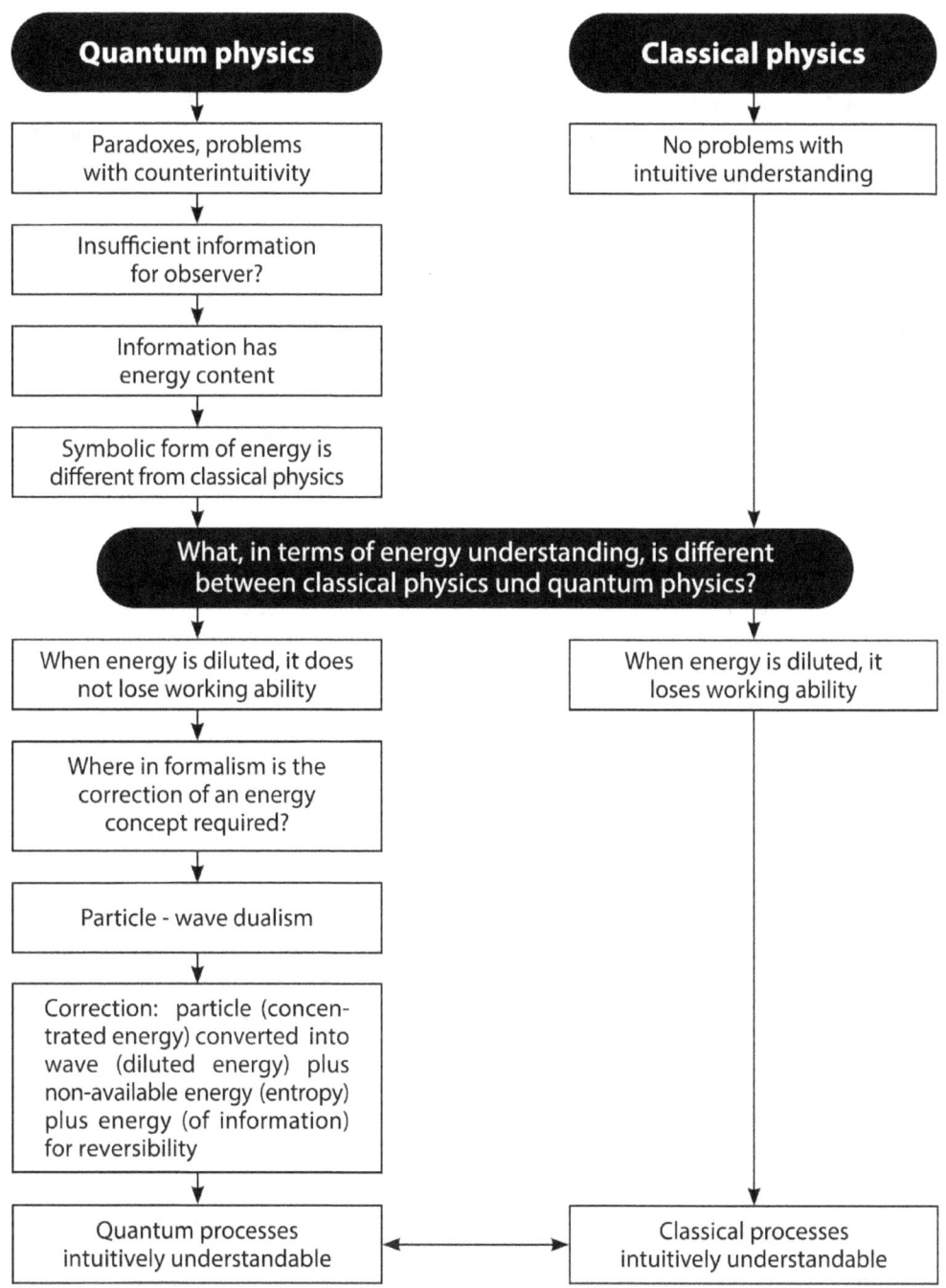

Fig. 41. Scheme explaining the strategy used to understand the kind of missing information in quantum mechanics as compared to classical mechanics moving towards an elimination of paradoxes.

The way this problem has been approached is summarized in the scheme depicted in fig. 41 and was discussed in chapter 13. Such a close analysis has shown that a particle cannot simply be equated with a wave. It was demonstrated that spreading out energy into space decreases its value of useful energy and information is needed to recreate a particle from a wave. Ernst Cassirer's considerations on the relevance of symbolic forms of thought for gathering cognition have helped us to address the problems of irrationalities and quantum paradoxes. The energy concept, which is used as symbolic form for understanding physics, is apparently different in quantum physics and classical physics (fig. 6).

By comparing energy statements both from classical and quantum mechanics it was realised that they are indeed contradictory in one important detail. The loss of working ability, when energy is diluted in space, is considered in classical physics (e.g. via the 2nd law of thermodynamics expressed in terms of energy and reflected in the behaviour of electromagnetic and acoustic energy fields), but is not considered in quantum physics. Statements on energy are definitely symbolic forms. Contradicting symbolic forms, however, cannot lead to consistent and rational perception. But interestingly, Ernst Cassirer himself did, at his time, not see a problem with quantum theory. On the contrary, he came to the conclusion that the complexity of "quantum understanding" was progress towards an ideal way of involving symbolic forms for scientific perception. He supported this understanding by substituting substance (mass, related to energy via $E = mc^2$) for function (e.g. general covariance, which looks at how random variables change together) as symbolic a-priori form. He interpreted paradoxes and other complications in quantum theory only as symptoms of a still incomplete transition towards a new perception (Cassirer, 1921). However, energy and its properties appear to be too fundamental and too important to be cast aside as a symbolic form.

I feel that Cassirer's concept of perceiving and understanding science is convincing, but his interpretation of quantum paradoxes is misleading and not thought out to the end. I believe that like other philosophers and intellectuals of his time, he apparently gave in to the conclusions from physicists who claimed to have monitored a real counterintuitive quantum world and world of relativity. The well known writer Hermann Hesse may be another example: He let the hero Vaseduva from his opus Siddhartha, which appeared in 1922, ask: "Did you learn from the river the secret that there is no time?" And the answer ended with "Nothing was, nothing will be, everything is, all is essential and presence", Herman Hesse was fascinated by timelessness and the absence of causality and other concepts, which could have reached him only via the then much discussed relativity and quantum theories. There may simply not have been enough knowledge for constructive criticism and contradiction.

A functional property of energy and its response to dilution in space as expressed via formula (11), is here invoked as a new symbolic form related to energy. As we have seen, it indeed facilitates cognition and intuitive understanding of quantum phenomena. However, since quantum states themselves do not involve energy conversion, formula (11) also considers the necessary energy of information for the recovery of the wave energy into its concentrated form as particle. In this way the overall energy balance is maintained. The information self-image functions like a Maxwell demon, acting from outside, in re-establishing a particle from a distributed wave. By introducing such a dynamic energy property into quantum theory via a modified particle-wave duality (formula (11)), the required additional information (as explained in fig. 19b) for eliminating the quantum paradoxes is provided. It is also understood why concentrated "dynamic" energy of a particle converts into diluted wave energy in an effort to decrease the energy per state. All together a minimum amount of energy is maintained, including the energy of information involved.

As demonstrated through the double-slit diffraction experiment, with the problem of quantization, the explanation of quantum correlation, and the interpretation of constant light velocity, the physical phenomena can now be rationally understood. The assumed activation of information, as compensation for entropy formation during reversible dilution of energy into space, turned out to be a key towards elimination of irrationality in quantum physics. It is a kind of an information self-image of matter. It can be considered to be exactly the information which had been suppressed in conventional quantum theory for the observer. When introduced into quantum concepts, it now helps the observer to avoid counter-intuition and irrationality. This additional information makes experiments understandable without questioning the experimental facts. In the double-slit experiment, for example, the information provided by the wave on boundaries within the second slit contained in the "negentropic" energy E_n, the energy of the information self-image, in formula (11) is reconverted into the energy of the particle, E_p. Thereby it may impose the recovered information related to the second slit onto the particle again when leaving the first slit (fig. 23). Individual particles passing the slits do not any more have to be present simultaneously within two slits to form a diffraction pattern. They rather monitor the second slit by continuously and dynamically changing from the particle form into the wave form via formula (11). Thereby individual particles, while passing only through one slit, "know" of the second slit (because they receive the information from it through E_n, the energy related to the information self image of the particle, which is "experiencing" the second slit).

Since this concept of a dynamic energy works as a symbolic form towards cognition in the classical sense, it is proposed to have the significance of a basic law for energy. The validity of such a law is supported by the demonstrated elimination of relevant paradoxes that made cognition in the classical sense impossible, but also by the expected consistency with experimental results (Tributsch, 2006, 2008). Classical quantum theory confronted us with a world where space has no relevance for energy. Where space disappears objects can, of course, be simultaneously at two different positions, reality may become blurred, and quantum exchange phenomena can be instantaneous. But after the effect of space was introduced via equation (11) (generation of entropic energy E_e and information energy E_n to neutralize it) the paradoxes disappeared. This is a remarkably logic explanation resulting from the developed strategy. It is, by the way, an extremely interesting fact that the missing information searched for in quantum theory turned out to be actual information itself. One can only guess how complicated the information self-image of a particle must be. If such a particle is split up into two quantum correlated sister particles, this information obviously also has to be subdivided, changed and split up. The difficulty of getting this self-image of information properly restructured when joint conservation laws are involved is the reason for the sustained link between separating, but still quantum correlated, particles. They remain linked by a joint self-image of information, so that manipulation of one particle may immediately affect and change the sister particle. Some information remains linked over larger distances while the quantum correlated particles are separating (fig. 21). Such a sequence of events and the new meaning of quantum correlation remain entirely logic and make the "spooky action at a distance" intuitively understandable. There would not be any fundamental irrationality left in quantum processes. Particles simply split up, but new is the self-image of information which may continue to link the particles over a distance. Non-locality does not have to be claimed. The separating particles are linked like two persons coordinating their activities via a cellular phone. However, there is, of course, still much more to learn.

It is interesting now to remember that the famous Bell theorem states that no local theory employing hidden variables can reproduce the predictions of quantum theory. However, the "dynamic energy" approach introduced the information self-image of matter and succeeded in giving a rational explanation of the two-slit experiment without having to assume that a particle exists simultaneously on two sites. It also succeeded in giving a reasonable explanation for the quantum correlation phenomenon. How can this be understood? The two new elements which have been introduced and are different from classical quantum theory are a "dynamic" energy and energy in form of a self-image of information (E_n). The dynamic energy concept, applied to quantum systems, suggested an alternative presence of particle and wave, mediated by energy in form of information. It will not any more be necessary to teach students that such fundamental quantum processes in nature are irrational.

There is a critical test for the existence of an information self-image of matter. Since turnover of information involves energy turnover, the information self-image of particles should somehow be observable and measurable in nature. In fact this information-energy phenomenon should be present around matter. It should also be more concentrated the higher the concentration of matter and less concentrated further away from it. And it should also exist within matter and act through matter. Evidence was given that this can only be gravitation. It is the medium, which communicates and implements the information, it is the medium sustaining quantum correlation. Information, the information self-image of matter, thus becomes a fundamental phenomenon for understanding the natural world in rational terms. Since human society is presently experiencing a revolution in information technology, it is entirely credible that one is dealing with an important and realistic natural phenomenon. Human information technology is based on modulated electromagnetic waves and an elaborate variety of technical installations that make use of it. Nature is apparently using gravitation for information transfer and processing and we have to learn how matter deals and interacts with it.

It speaks for the consistency of this new quantum interpretation that additional paradoxes disappear. Since electrons in orbit around nuclei also aim at generating a wave, and since a minimum energy situation has to be approached, the energy of information is also minimized. This means that the simplest wave pattern possible only is accepted around a nucleus of an atom, because it requires a minimum of information. This wave pattern corresponds to integral numbers of full wavelengths around the atom. Such a condition enforces quantization. This way quantization becomes equally understandable. It is a consequence of minimizing information.

Also the claim of quantum physics that causality does not apply is invalidated. It is not non-causal statistics but deterministic chaos, which controls the decay of radioactive particles. The reason is that in a time-oriented energy world, elementary particles can be understood as products of self-organization of energy. Before a particle decays, it shifts into chaos due to a minimal interior parameter change, and decays according to a statistics determined by deterministic chaos. This deterministic probability cannot be distinguished from purely statistic probability. Experimental facts are still valid.

Similarly, non-locality in quantum physics can be discarded with the following argument: first, particles are no longer statistically delocalized along a wave. A wave is a product generated from a particle. Therefore, particles are no longer found simultaneously at distant locations. Experiments with entangled particles do not necessarily support non-locality either. The link which temporarily exists between separating quantum correlated particles is a link of information (fig. 21). As long as it exists particles could be considered as undivided. I compared them with two people coordinating their activities via a cellular phone. Such a reality can be understood classically. There is no need to conclude that their activity is non-local. The quantum paradox of non-locality equally disappears.

In contrast to Healey (2012), who proposes to ignore paradoxes and just to use quantum theory as a tool of physics, I was correcting and extending the energy concept in quantum physics (as explained in fig. 19b and via equation (11)) thereby succeeding in activating additional information towards improved cognition.

Fact is that the new interpretation of the quantum phenomena has become intuitively understandable. The newly defined particle-wave duality (relation (11)), considering an information self-image of matter, has made the quantum phenomena rationally transparent. I believe that this makes the quantum description also much more complete. An interesting result is also that the originally searched for missing information in quantum theory actually turned out to be information itself. And there is an additional interesting result: the description of quantum reality, of particles or energy, has to be complemented with information on this reality to become rational.

There is now promise of a much deeper insight into fundamental aspects of quantum processes. It was possible to drastically modify the interpretation of quantum mechanisms without challenging the experimental reality.

The potential consequences of the developed arguments for quantum physics are, in my opinion, significant. It will take some time to understand them all. But the situation could be handled with patience and determination: the induction problem, the problem of validity of such a newly claimed natural law, can be treated according to science philosopher Karl Popper`s proposal as a "partially decidable" statement on truth, which could be methodically tested via attempts towards falsification (Popper, 1997). This has a reason: only statements can be considered to be scientific for which it can be defined, under which conditions they could be refuted (falsified). Suggestions are given further below.

The expected advantage is that new frontiers of cognition may be attained there where paradoxes had suggested bizarre concepts, ranging from the double presence of objects, to energy generation out of nothing and non causal phenomena.

Putting together the main arguments, it is claimed that the one century-old effort by quantum physicists to convince the public of an irrational foundation of physics is based on the use of a partially faulty and incomplete mechanism for creating perception. The empirically derived quantum laws have yielded energy properties, used as symbolic forms, different from and partially contradictory to classical ones. When correcting them to the same energy properties (concerning the relevance of space for energy), counter-intuition is overcome and paradoxes disappear.

 A remarkable parallel achievement of the suggested concept is that the experimentally verified constancy of light velocity in all relative velocity frames is also intuitively understood. It is simply implemented by the information self-image of matter. It is the consequence of a continuous (dynamic) intermediate presence of the light particle in the form of information (E_n in relation (11)), which explains the recovery of an identical photon property E_p with given velocity. Constant light velocity is apparently fixed by fundamental law. If it were not fixed, an extremely fast object approaching a high speed light source from the opposite direction would be able to implement a super light velocity. The situation actually functioning is like recovering an identical information pattern (for television or a 3-dimensional printer) via digital signals in entirely different velocity frames (fig. 25). This is intuitively perfectly understandable.

Einstein declared the constant light velocity, which had previously been experimentally verified (Michelson experiments), simply a property of space and as a consequence postulated the four-dimensional space-time, which has now become the entirely accepted concept shaping our understanding of the universe. Based on the described straightforward, trivial local interpretation of the constancy of light velocity this is not true. Within the theory of relativity this unification of space with time, where space and time coordinates are mixed in a four-dimensional structure, had been identified to be the only possible consequence of the experimentally verified paradox of an always constant light velocity. If, however, the constancy of light velocity can be explained locally, as a property of quanta constantly reassembled to create always identical properties from information, this would not require a four-dimensional space. One does not need a space which has to enforce constant light velocity. The supposed curvature of space-time through energy-carrying obstacles, including a "curved" time, a time which is being manipulated, would in this case be replaced by a gravitation effect of these energy obstacles on the new defined quantum state defined in relation (11).

This conclusion is of remarkable philosophical importance. The dynamic energy approach offers a significantly simpler and more convincing "symbolic form" compared to four dimensional space-time and it provides a new interpretation of gravitation. Imagine an "empty" space with all the complicated properties of implementing movements, of manipulating time (which in addition is supposed to be an illusion) and of accelerating objects. The possibility of doing without a four-dimensional space-time is, for the moment, a highly surprising conclusion since it was unexpected. However, the four- dimensional space is also counterintuitive as are several additional mechanisms linked to it. In addition, it was shown above that statements of the special and general theory of relativity vary significantly with an assumed changed light velocity for signal transmission. An ideal, infinitely fast signal transmission, for example, which guarantees simultaneity all over space, eliminates dilation in time and length. One is dealing with reality manipulated by signal transmission. Do we want to understand the universe as a universe manipulated by the velocity of light, by signal transmission? Or do we want to understand it as it is, imagining an ideal mechanism of information transfer?

The dynamic energy properties introduced as new symbolic forms to better understand diverse physical processes are basically simple, justified via the fundamental principle of least action, and they are logical. The energy concept is only changed from a "sleeping" energy to a "dynamic energy". Nevertheless this is a paradigm change since it introduces a time arrow into physics. In fact, it is an arrow of elements of action (energy multiplied by time), but a time arrow can be deduced or approached by dividing action by the energy turned over. This is done in clocks and in our brain. But this simple change of energy concepts can provide a new approach for the interpretation of various phenomena, ranging from time (a track or consequence of energy turnover) to space dynamics and consciousness (self-organization of information via a time driving energy). An energy-converting world, in which time orientation is fundamental and energy tends to decrease its presence per state, can self organize. The reason is that there is a "before" and an "after", circumstances needed for feedback processes to occur. Our present perception of a physical world, with fundamental processes being time invertible, and time being an illusion or just a statistical path towards entropy increase, cannot properly explain time-oriented reality. When this is nevertheless attempted, theories are based on audacious mathematical interpretations. The concept developed here leads us to expect that when energy is turned over, natural processes occur based on the principle of least action. Action (energy multiplied by time) is turned over, which gradually changes the appearance of our environment. Our brain measures this flow of elements of action and divides it by the energy turned over or an assumed constant energy quantity and calibrates it. The result is the experience of time. Our watches also function in a similar way. They obtain energy which they release in small action quantities that advance the experience of time.

The proposed and modified energy properties also allow starlight, which spreads in space, to generate entropy. This way it can respect thermodynamic law which requires entropy production from spreading light. Here again, conventional quantum physics neglects the effect of space on energy. In contrast, in the corrected approach photons can generate entropy. The accumulated entropy in space is recognized to be the microwave background radiation. The Big Bang theory as well as the process of inflation of empty space are consequently questioned. This also concerns the near equilibrium interpretation of black holes and quasars. The gigantic radiation flux from quasars coupled to black holes is interpreted as a phenomenon of far from equilibrium maximum turnover of entropy.

A really astonishing result was also the realization that the information mediating between particle and wave is gravitation, a phenomenon which is still little understood by modern science. This is an entirely different explanation offered compared with that of the general theory of relativity. However, the latter cannot say what gravitation really is, while the dynamic energy approach clearly identifies it as energy related to an information self-image of matter which is thus quite well characterized and deals with the particle-wave dualism. One can consequently draw the link between the large dimensional effects of gravitation in space and its role in quantum processes. There is also an additional interesting consequence. The role of information as an active link towards reversibility between concentrated, active energy and expanded, degraded energy does not only function in quantum mechanical processes. It may, as gravitation in space, and since gravitation is so important to the concept of space dynamics, also mediate the interrelation between a primordial and a worn out universe (fig. 37). This is an indication for a self-similarity existing between the function of the sub-microscopic quantum world and the universe (fig. 38). This self-similarity again would reveal the necessity to assume a fundamentally dynamic energy and an oriented time for understanding our world. In addition there would be a straightforward link between the quantum world and the function of the universe. This concept would also suggest a universe that can cyclically renew its potential to perform as we know it now. The information self-image in the form of gravitation distributed in space will transform the worn-out universe and recreate its primordial state. It is a return from a "chaotic" universe to an ordered one. It is like a return from chaos to order via a change of some parameters and the involvement of energy, which has been set aside as an information self-image. A mechanism of this type, a consequence of the dynamic energy concept, would not contradict scientific experience, which man has accumulated up to now.

Such a universe, the "self-image universe", which periodically renews itself to allow matter and information to self-organize towards life, mind and intelligence and to grow to a high degree of order and perfection, is intuitively more convincing than a Big Bang universe starting from nothing in chaos, dominated by time invertible physics, tolerating man as a chance product and ending dead and in darkness. It also allows more room for philosophical and religious reflections on our existence and depicts nature as more conciliatory and eternally functioning. This is nature which man can attempt to understand and to tolerate.

How the idea of "dynamic" energy, which was mathematically derived from the principle of least action (chapter 10), has helped to eliminate irrationalities and paradoxes in different areas of science and has yielded new concepts and theories is shown as an overview in fig. 42. The irrationalities, which have been eliminated, are marked. The role of information and its involvement in quantum and space mechanisms is emphasized.

Let us, in this context, address quite an important point – that of the more general validity of the advanced few basic arguments. A relatively simple correction of the understanding of energy has been implemented, which has been shown to be derivable from the principle of least action (fig. 16). For quantum physics this has been expressed in relation (11) under consideration of the effect of space on energy. It has eliminated the non-locality problem (double-slit experiment, fig. 23)(paradox 1) and in a simple way has explained quantum correlation (fig. 28)(paradox 2). It has also resolved the constant light velocity paradox (paradox 3). Quantization became intuitively understandable as by infor-

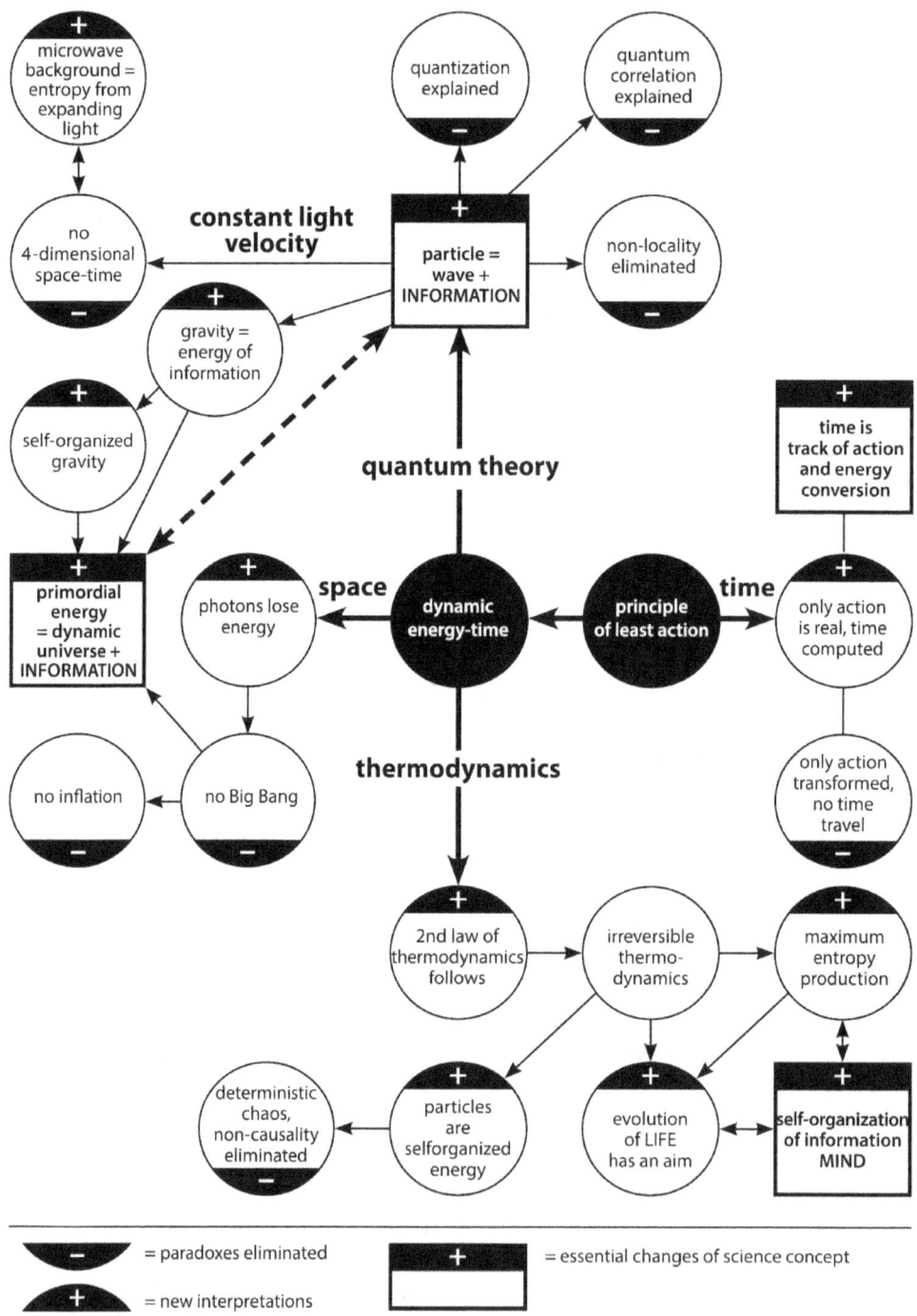

Fig. 42. Overview explaining how the "dynamic" energy concept, which can be derived from the principle of least action, is reshaping areas from quantum physics and thermodynamics to space science and concepts of time. The areas marked are those where paradoxes are eliminated or new interpretations are provided.

mation on a particle wave duality of minimum energy content (paradox 4) and non-causality was replaced by deterministic chaotic statistics (paradox 5). The approach has also eliminated the need for a four-dimensional space-time as a consequence of an ever constant light velocity (paradox 6). Since energy is now defined to be time oriented and thus related to time, only action, energy multiplied by time, is real and can be transformed.   Also the paradox of time travel is eliminated since action is invariant during transformation (paradox 7). The identification of gravitation as the information mediating the interaction between concentrated and diluted energy, and between primordial space and the entropy-rich final space led to fertile new theories and an entirely new concept of the role of gravitation. The Big Bang mechanism (paradox 8) (including inflation of space (paradox 9) in the presently discussed form is excluded as a realistic possibility. The microwave background is identified with chaotic energy, or energy related to entropy, generated by expanding starlight and self-organization activities of galaxies. The super-gravitation effect of invisible dark matter (paradox 10) was differently and rationally explained and the existence of dark energy and an exploding universe questioned. These appear to be quite astonishing results considering the fact that basically only two essential changes have been introduced. One is the claim of a time-oriented, dynamic energy. It has mathematically been derived from the principle of least action, which is fully recognized in science as a most fundamental principle of nature. Its dynamic interpretation is considered to be equivalent to the "dynamic energy" statement. The second is a correction of an error, which quantum physicists introduced into quantum theory. It is a false assumption that diluted and concentrated energy are equivalent in particle and wave descriptions. This simply ignores the relevance of space for energy. These corrections have eliminated quantum paradoxes and also questioned the evidence for an explosive expansion of the universe.

Such remarkable results in resolving paradoxes and avoiding counter-intuitive mathematical speculations seem to provide a sufficient basis for challenging the claim of some scientists that the fundaments of nature are irrational (Zeilinger, 1999, 2000). It is concluded here that the presently widely accepted formalism for describing fundamental nature is incomplete. The ease with which the small modifications introduced here have allowed us to restructure essential foundations to understand nature and to eliminate important irrationalities and paradoxes is an additional convincing support for the assumed strategy. The simplicity with which also the difficult challenge of gravitation could be handled, how it could be linked with quantum processes and how a new approach to evolution was found is an additional evidence for fundamental consistency.

After all, it is well known that the Franciscan friar William of Ockham from the 14th century suggested, that "one should not admit more causes of natural things than is sufficient to explain them". This rule of thumb for scientists, also known as Ockham`s razor, would clearly favour the "dynamic" energy approach presented here against the irrational theories for diverse important natural phenomena, which have been criticised and replaced. In addition, describing nature in terms of paradoxes is not the same as explaining it rationally.

The dynamic interpretation of energy and its effect on quantum physics as well as on the theory of relativity and on space models motivates a re-evaluation of established concepts. The chance to return to a reasonably logical worldview and science foundation seems to justify additional efforts. But first, of course, the presented concept will have to be confronted by scientists who are in the meantime convinced that essential foundations of physics are not rational. My theoretical approach suggests some straightforward, simple, but crucial experimental tests to contradict and to falsify (disprove) its relevance: in order to challenge the concept of "dynamic energy" it should be tried to demonstrate that a reasonably complex energy converting process can be entirely inverted in time. This would underline the existence of time-invertible processes. Critical scientists should also try to demonstrate that energy retains its full working ability when extensively diluted in space. If they succeed in doing so (thereby, however, also contradicting the empirical 2nd law of thermodynamics) our hypothesis is wrong. These are statements of falsification of the proposed hypothesis, which makes it a scientific theory (Popper, 1978).

It should also be investigated where and how spreading stellar radiation deposits entropy in the form of non-available, chaotic energy. Such a mechanism is unavoidable on the basis of thermodynamics, but is difficult to realize with present quantum mechanical concepts. Energy spreading into space is, as we have already seen, a critical issue related to irrationality and paradoxes in quantum physics. Another important test could relate to my argument that information cannot simply be omitted in mathematical statistical calculations of entropy, as was done by Boltzmann and later within "coarse graining" approaches for calculating the phenomenon of entropy increase. If this argument is confirmed, why does time then develop in one direction? Time orientation must then be fundamental, generated by chains of action, as proposed here. The new interpretation of gravitation as an information self-image of matter is also a relevant statement, which could be critically reviewed. It opens new opportunities to understand this mysterious phenomenon, which penetrates all matter, cannot be shielded and knows no opposing phenomenon, no anti-gravitation. But with the notion that gravitation is in fact a fundamental type of information, the great challenge for the future will be to read and decode this information and to understand how it interacts with matter.

Information as a balancing factor between concentrated energy-rich and diluted, energy-poor, entropic form of matter was recognized as fundamental for quantum systems. It may also be relevant for the macroscopic function of the universe. Such a conclusion is supported by the fact that gravitation, identified with the information self-image of matter, is really an important control factor for the universe. The sub-microscopic world would be subject to similar principles as the macroscopic. The universe could thus essentially be based on self-similar principles, which itself speaks for energy and time as dynamic quantities.

Another critical test could be the derivation of the "dynamic energy" concept from the principle of least action (chapter 10). I consider my mathematical arguments to be correct and consistent. If this is agreed, how can one then assume that energy has no orientation but is "sleeping", just acting as a number? There is another interesting argument. Those established theories, quantum theory, theory of relativity, Big Bang and inflation theory, which are based on or generate irrationalities, pay no attention to the potential of information as a basis for astonishing mechanisms. When these theories were born, the potential of information mechanisms was not yet evident, but our modern world is witnessing a revolution of information technologies. Why should fundamental processes in nature not have taken advantage of these amazing possibilities which information offers on the basis of obvious natural laws? In the "dynamic energy" approach presented here information turned out to be fundamental and a key to overcoming irrationalities and paradoxes. This adds credibility to the presented ideas and students who live in the current information age will understand.

An important result is also the surprising conclusion on the path of evolution. It aims at maximum energy turnover within the restraints given. This coincides with the entropy law identified for self-organized systems far from equilibrium. Life, ecosystems and human societies are such self-organized systems. Here, maximum entropy turnover is expected within the restraints given. This also requires maximum energy turnover. Starting with self-organized living systems, the inevitability of evolution towards maximum energy and entropy turnover can be assessed in a simple thought model (fig. 43). Let us first imagine self-organized living systems or inorganic dynamic systems competing for maximum entropy production, within their restraints, as explained in chapter 11. We are looking back to the time of early man, with wild cows and birds around in the wilderness, with a natural forest and a river which can freely select its meandering path. Early man has evolved to represent mankind of today and the forest has turned into an agricultural landscape and the river into a canalized energy- producing waterway. The cattle were domesticated for meat and milk production, the bird population for egg production. Man's energy and entropy turnover is also evident due to big cities, high traffic, large industrialized complexes, the exploitation of fossil energy and the liberation of greenhouse gases. Also the strongly increasing turnover of

information contributes to energy consumption, which, from the time of primitive man has increased 100-fold. The self-organized system "man" evolved exploiting and enslaving many of the former co-existing self-organized systems and taking advantage of additional energy sources from the earth. Man's energy conversion activities have profoundly changed the landscapes of the world. One recognizes the path of evolution and also the progress of time as the track of progressing and increasing energy turnover. Without an increasing energy conversion rate for modern man, there would not have been such a dramatic change.

competition of systems with maximum entropy production

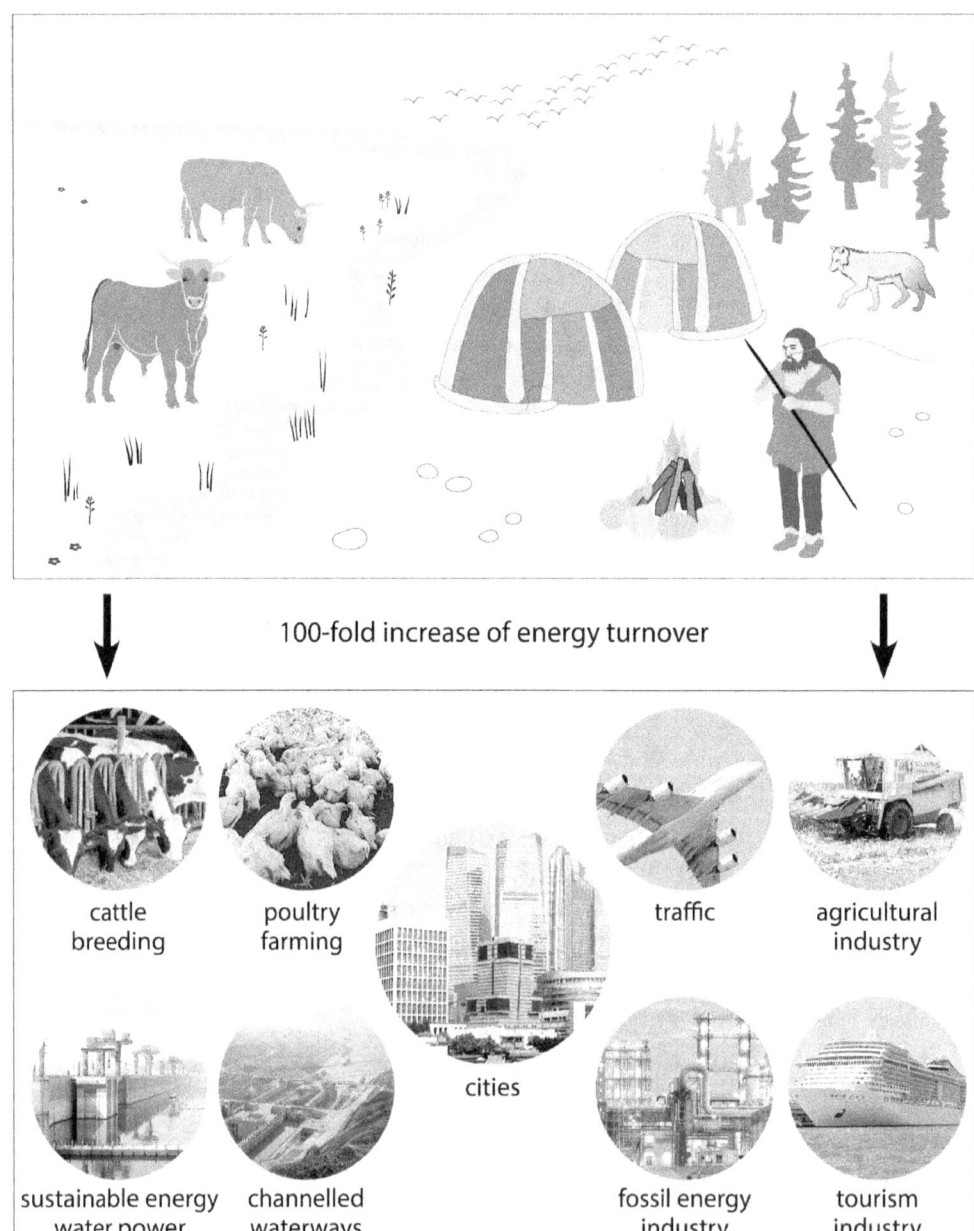

100-fold increase of energy turnover

cattle breeding

poultry farming

cities

traffic

agricultural industry

sustainable energy water power

channelled waterways

fossil energy industry

tourism industry

maximum energy / entropy turnover results

Fig. 43. When comparing the competition of self-organized systems during the time of early man with the present activities of man, one realizes the trend of increasing energy turnover. Also the effect of timeflow and evolution is recognized. It is visible as a track of energy turnover, which has profoundly modified the appearance of landscapes.

Recognizing such an evolution law of maximum energy turnover for mankind is important, but disturbing. No doubt it supports the impression, which critical observers get from the state and evolution of our environment. By generating too much entropy we are seriously harming our environment. It will require profound studies and intelligence to find out how mankind can cope with such a situation. However, there is a fascinating alternative interpretation of evolution's possible destiny. A drive towards maximum energy and entropy turnover has been a precondition for the self-organization of information and evolution of mind (fig. 27). Information systems have to be pushed far enough from equilibrium for self-organization, and this requires a considerable through-flux of energy. We should recall that our brain with a weight of only 1.3 kg consumes one quarter of a person's energy supply. Is the generation of conditions of high energy turnover for enabling mind and intelligence actually the aim of evolution? Is the enrichment of the universe with highly developed civilizations, which take advantage of consciousness, intelligence and mind, the actual aim of evolution? One has, in fact, two different interpretation possibilities for the aim of evolution. One is negative and dramatic: the self-organized living system that will dominate its environment while harvesting maximum energy, and which will ultimately destroy it. The other is positive: intelligence and mind will help the living system to understand its situation and to act against the gradual accumulation of entropy, which will deteriorate its ecosystem.

Evolution of intelligence and mind would be an aim, which could be understood as positive, even attractive. In this case, intelligence may then also be the tool provided and be available to civilizations for coping with the burden of side effects from too high an energy consumption. Such considerations about an aim in evolution have yielded surprising new options for the human imagination. There is not only room for more profound scientific considerations, but also for religious-philosophical thoughts. The gap between down-to-earth scientific conclusions and expectations from communities counting on divine creation has become more narrow. This is an interesting, unexpected new perspective, which the model of the "self-image universe" offers. It somehow psychologically encourages man to perform better and with more vision, to cooperate in a fascinating project called evolution. Intelligence and mind as the aim of evolution appears to open the door to a mystic universe. Is this something negative? This reminds me of Erwin Schrödinger, who, when reflecting on the then profound mysteries of quantum processes commented: "A purely rational world view entirely without mysticism is absurd". I have tried to bring rationality into basic science and quantum physics and found new mysticism on another frontier: mind as the aim of evolution?

Let us return to the final assessment of this exploration of irrationalities in science concepts. It should be emphasized that a scientific model, which can avoid paradoxes and irrationalities, has already demonstrated it is better at explaining reality. This is a philosophical claim based on the assumption that nature functions on a logic basis and our perception has evolved to understand this. I would like to come to a concluding remark and answer the question posed in the title of the book:

"Nature is rational, because it works!"

And indeed no irrational mechanisms have been identified within the wealth of engineering and technology accomplishments in living nature. Is an entirely rational technology somehow to be expected on the basis of irrational fundamental laws (fig. 44)? Science should definitively return to a rational description of nature. There is a reason why evolution has shaped man with the profound feeling for causality and rational relationships in nature. His survival depends upon it.

According to the ideas promoted in this book, the universe is subject to irreversible mechanisms. It is driven by energy, which generates action and thus experiences changes, which our brain or clocks can recalculate and register as time. This time is a substitute and handling parameter for action. Time itself has no substance. It cannot be measured directly and cannot be imposed to modify the environment, but it is an indication and a calibrated measure for on-going action. Self-organization processes are a natural consequence of time orientation, because feed-back mechanisms are permitted. They consequently dominate evolution by shaping galaxies and planets as well as biological life forms. The mechanisms by which energy is turned over from useful to a non-useful, chaotic form follow straightforward regularity. It is for this reason that our world is logical and we observe causality. Our brain, which is subject to the same kind of processes, recognizes natural mechanisms as rational. I believe that a rational basis of understanding involves a minimum turnover of information (which, as we know, is related to energy), once the content is understood. The brain is satisfied, which may be an adaptation from evolution. In practical life, logic situations turned out to be less harmful, because circumstances could be understood. In contrast, an irrational knowledge structure keeps significantly more information and non-answered questions activated. Therefore, since an understanding is not reached, no minimum in mental energy turnover is to be expected and practical consequences may be destructive. Brain activities are not balanced in this case.

Most importantly, the effort to understand our physical world on the basis of a dynamic energy, which leads to rationality, also conduced to an integration of consciousness and mind into the building concept of nature. This is a remarkable achievement and a further touchstone for the dynamic energy approach. By allowing information, as a quantity related to energy, to self-organize, it can fill a gap which existing theories have not yet been able to bridge. It opens the potential of including those phenomena and abilities, which enabled man to understand his position in nature, and to explore the laws which govern it. It was the realization that evolution follows an aim in maximizing energy turnover,

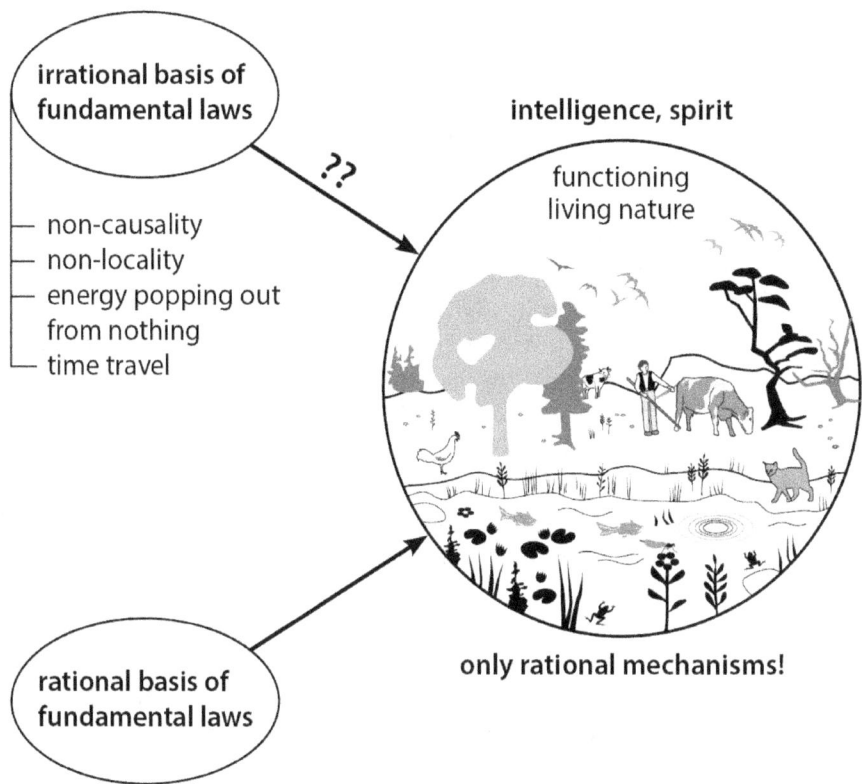

irrational basis of
fundamental laws

intelligence, spirit

??

functioning
living nature

— non-causality
— non-locality
— energy popping out
  from nothing
— time travel

only rational mechanisms!

rational basis of
fundamental laws

Fig. 44. The wealth of biological creativity can be explained on a purely rational basis and man has an instinct for rationality and causality. This would not be credible on the basis of an irrational foundation of nature. We should deduce from the rational function of the complex living nature that we exist in a rationally founded and functioning universe. Claims of irrationality in nature are claims of ignorance. This can, for example, be deduced from the explanation of quantum correlation presented here, once named "spooky action at a distance" (fig. 21). It became rationally understandable, although it still remains a puzzling phenomenon.

which conduced to the surprising finding that the actual aim of evolution may be the facilitation of mind. Consciousness and intelligence require such high energy turnover and selected living systems gain the opportunity to become conscious and intelligent.

This re-evaluation of basic features of nature`s function offers the potential of a deeper understanding of our world and our existence. Answers to profound questions from quantum physics to gravitation, from evolution and cosmology turned out to be different from conventional ones. The search for a science structure free of irrationalities and paradoxes has thus opened the door to new philosophical challenges: Is it not reasonable to explore the here proposed alternative gravitation theory, when experimental proof of dark matter cannot be established? Why is the development of spirit an aim of evolution?

The path towards the truth, however, has to confront some established theories, which I claim, are wrong. Profound discussions will be necessary. A main topic is the question of admissibility of irrationality in science. And there is an important point resulting from these studies: entirely different models for the reality of our world and the universe can be derived from logic understanding when irrational theories are abandoned. It is claimed that the rational ones are the correct models, at least they offer a much higher chance to be correct.

There is another matter which should concern us: science should again learn to occasionally admit that it cannot explain everything. In the model of our universe discussed here such a mystery would, for example, be the way the information self-image, identified with gravitation, would restore particles or even the primordial world. On the basis of superficial scientific reasoning it would work as information can do that, but details appear to remain out of reach for our present understanding. The German poet and writer Goethe was also quite dedicated to private scientific research and at the same time maintained quite a healthy measure of self-criticism. He said that the highest fortune of thinking man is to have explored what can be explored and to admire what cannot be explored. It is indeed much more reasonable and human to admit that one does not understand, than to claim and insist that reality is irrational, because this is a comfortable excuse. Then scientists will also try harder to find rational solutions for the problems and challenges of our world.

37 Concluding remarks

After guiding the reader through this alternative world of science, I have to admit something. I was never able to discuss my "dynamic energy" hypothesis and the "self-image universe" with other scientists from the fields of physics, chemistry or biology. I know that this is not good procedure. Scientists should try to sharpen their power of judgement in discussions. I am, however, not to blame for this. Since my approach was quite trans-disciplinary, close colleagues, typically working in the field of physical chemistry, were too specialized to be interested in the questions addressed in this book. Or they were not sufficiently informed to address my questions. I had friendly relations to one scientist, who was a quantum physicist. When I approached him with my ideas, he was shocked. He said he could not possibly get involved and justified it by saying " I make my living with quantum physics". So, I attempted to get the necessary response from the science community by submitting publications to relevant journals. This did not work. Over a period of two decades I tried approximately twenty times to submit papers related to my considerations on energy and information. They were all rejected, with exception of a small contribution in a little read journal. In most cases the manuscript was not even forwarded to referees and was returned without scientific criticism. In one case, I remember, the comment was "quantum theory has proven to be correct and complete". Physicists have apparently learned to live with irrationality and felt comfortable. So I changed my strategy. I thought, that philosophers should be interested in getting rid of non-logic, irrational concepts in science. I tried to adapt my papers to the style and vocabulary of philosophy and submitted them five times in vain to well established philosophical journals. My papers were rejected. In part, no comment was given, or there was not sufficient relevance for philosophy. One comment shocked me: "it is just not true that the constancy of the velocity of light is unintelligible". How can one possibly comprehend that light emitted from high speed sources when approaching and when leaving can have exactly the same velocity. I realized that philosophers have also embraced irrationality. Of course I was not happy about so much effort and work in vain, but I did not attribute the difficulties I went through to personal inexperience. I had published many scientific publications and felt competent in research, related to energy. I gradually found comfort in the consideration that those who insist on irrational interpretations of nature can never expect to reasonably understand it. As scientists, they are punishing themselves. A science loaded with irrationalities is not real science, but gives only an illusion of science, supporting dubious careers. With this book I want to address, above all, people interested in nature, the destiny of our world and the universe, people who are not happy living with irrationalities and who care for nature and science because of the fascination they hold. The possibility of getting relevant scientific information via the Internet and via documentaries on television has enabled many people without an opportunity to graduate or without careers in science to reach a quite good standard of understanding. I want to reach them and to interest them. I would also like to encourage students to rely less on abstract mathematical theories, and to concentrate more on rational explanations of our world. Our society has to stay rational in order to solve its increasing problems. I also hope that my critical approach to irrationality will finally spare students from having to digest incomprehensible interpretations of quantum and relativity phenomena. For me, the concept published here is my way of understanding our world and the universe. It is rational and I feel sufficiently comfortable with it. In German the word "scientist" is "Wissenschaftler", someone who creates ("schafft") knowledge, "Wissen". Irrationalities and paradoxes are actually not knowledge, but rather information on "not understanding". My effort to challenge irrationality in science was simply part of my mission and responsibility, and a late answer to the questions left from my time as a student. The science community has apparently already got used to irrationalities and has at present little time to listen. This is the reason for this book being written.

References

Aharonov, Y., Rohrlich, D., 2005, Quantum Paradoxes, Quantum Theory for the Perplexed. ISBN-13: 978-3-527-40391-2 Wiley-VCH

Amelino-Camelia, G., 2000, "Quantum theory's last challenge" Nature 408, 661- 664

Arp, H.C. ,1987, Quasars, Redshifts and Controversies. Inter-stellar Media.

Antoniou, I. E., Prigogine, I., 1993, Intrinsic irreversibility and integrability of dynamics, Physica A 192, 443.

Aspect, A., et al.,1981 *Experimental Tests of Realistic Local Theories via Bell's Theorem*, Phys. Rev. Lett. 47, 460

Aspect, A., Grangier, P. and Roger, G. (1982) *Experimental Realization of Einstein-Podolsky-Rosen-Bohm Gedankenexperiment: A New Violation of Bell's Inequalities*, Physical Review Letters, Vol. 49, Iss. 2, pp.91-94 doi: 10.1103/PhysRevLett.49.91

Aspect, A., Dalibard, J., Roger G.,1982, *Experimental Test of Bell's Inequalities Using Time-Varying Analyzers*, Physical Review Letters, Vol. 49, Iss. 25, pp. 1804-1807

Baumgartner, M., Regularity theories reassessed, http://philsci-archive.pitt.edu/2874/1/regul_recon.pdf , retrieved 06/04/2014)

Bell, J. S., 1964, On the Einstein-Podolsky-Rosen Paradox, Physics 1, 195-200.

Berkovitz, J.,1998. a) Aspects of Quantum Non-Locality I: Superluminal Signalling, Action-at-a-Distance, Non-Separability and Holism. *Studies in History and Philosophy of Science Part B* 29 (2):183-222

Berkovitz, J. ,1998,. b) Aspects of Quantum Non-Locality II: Superluminal Causation and Relativity. *Studies in History and Philosophy of Science Part B* 29 (4): 509-545

Berliner, L. M., 1992, Statistics, Probability and Chaos: Statistical Science, Vol. 7, No. 1, pp. 69-90

Black hole, 2014: http://en.wikipedia.org/wiki/Black_hole, retrieved 04/05/2014

Bohm, D ; Hiley, B. 1993 , The Undivided Universe, London , Routledge

Brillouin, L., 1951, Maxwells demon cannot operate. Information and entropy I, J.Appl. Phys. 22, 334-337

Bub, J., 1979. Some Reflections on Quantum Logic and Schrödinger's Cat. *British Journal for the Philosophy of Science* 30 (1):27-39

Butterfield, J.,1995. Quantum Theory and the Mind. *Proceedings of the Aristotelian Society* 69 (69):113-158

Capra, F.,1979, The Tao of Physics: An Exploration between the Parallels between modern Physics and Eastern mystics. Boulder Shambala

Cassirer, E. Philosophie der symbolischen Formen, Die Sprache (1923), Bd. 1; Das mythische Denken (1924), Bd.2; Phänomenologie der Erkenntnis (1929), Bd. 3; Darmstadt (1997)

Cassirer, E., 1964, Zur modernen Physik. Darmstadt

Cassirer, E. a) Determinismus und indeterminismus in der modernen Physik, in E. Cassirer: Zur Modernen Physik, Oxford (1957), 127-379 (first print: Göteborg, (1937); b) Zur Einsteinschen Relativitätstheorie, Berlin, (1921)

Cattani, C., 2010, Mathematical Problems in Engineering,Volume 2010, Article ID 507056, http://dx.doi.org/10.1155/2010/507056

Chronology of the universe, 2014, http://en.wikipedia.org/wiki/Chronology_of_the_universe; retrieved 04/02/2014)

Clauser, J.F., and Shimony, A., 1978, *Bell's theorem: experimental tests and implications*, Reports on Progress in Physics 41, 1881

Clauser, J., and Freedman, St., 1972, Experimental Test of Local Hidden-Variable Theories Phys. Rev. Lett. 28, 938, DOI: http://dx.doi.org/10.1103/PhysRevLett.28.938

Conservapedia, 2014; www.conservapedia.com/Theory_of_relativity

Collins, J., Hall, N., and Paul, L.A., editors: 2004, Causation and Counterfactuals, Cambridge: MIT Press

Daecke, S.M., 1991, Worte in Zeit und Raum, p. 124

David-Néel, A.,1994, Mein Weg durch Himmel und Hölle, Scherz-Verlag, (original edition: Voyage d'une Parisienna à Lhasa, Librairie Plon, Paris)

Dickson, W.M.,1998. Quantum Chance and Non-Locality: Probability and Non-Locality in the Interpretations of Quantum Mechanics. Cambridge University Press

Einstein, A. 1905, Concerning an Heuristic Point of View Toward the Emission and Transformation of Light, translated in: American Journal of Physics, 33, n. 5 (1965)

Einstein, A., 1911, Die Relativitäts-Theorie, In: *Naturforschende Gesellschaft, Zürich, Vierteljahresschrift.* 56, 1911, S. 1–14.

Einstein, A.,1920, Ether and Theory of Relativity, Lecture given at the University of Leiden on May 5th, 1920.

Einstein, A., Podolsky, B, Rosen, N.,1935,: *Can quantum-mechanical description of physical reality be considered complete?,* Phys. Rev. 47 (1935), p. 777 - 780, doi:10.1103/PhysRev.47.777

Feynman lectures on Physics, 1964, Feynman, R.P., Leighton, R.B., Sands, M., volume II, chapter 19, page 8), Addison Wesley

Fractal antenna, 2010; http://en.wikipedia.org/wiki/Fractal_antenna;

Framingham Heart Study 2012, Driver, J.A., Beiser, A., Au, R.,Kreger, B.E., Spiansky, G.L. Kurth, T., Kiel, P.D., Lu, K. P., Seshadri, S., Wolf, P.A. Inverse Association Between Cancer and Alzheimer`s Disease: Results from the Framingham Heart Study; British Medical Journal, 2012; 344:e1442 doi: 10.1136/bmj.e1442 http://www.medscape.com/viewarticle/760142

French, S., 2001, Symmetry, structure and the constitution of objects, http://philsci-archive.pitt.edu/327/

Galaxies, 2013, http://www.dailygalaxy.com/my_weblog/2013/06/500-billion-a-universe-of-galaxies-some-older-than-milky-way.html, retrieved 04/02/2014

General relativity, 2014, http://en.wikipedia.org/wiki/Mathematics_of_general_relativity

Gödel, K., Köhler, E., Buldt, B., 2003,: *Kurt Gödel: Wahrheit & Beweisbarkeit*, Verlag Öbv & Hpt, 2003, ISBN 3209038341

GPS, 2014, http://www.alternativephysics.org/book/GPSmythology.htm

Gravity probe B, 2014; http://physics.aps.org/articles/v4/43

Gregg, B.A., Nozik, A.J., 1993, Existence of a light intensity threshold for photo conversion processes, J. Phys. Chem. [1993],97,13441-13443

Gut, B. J.,1981: Immanent-logische Kritik der Relativitätstheorie. Oberwil b. Zug: Kugler, 1981. 151 S.

Hall, N., 2004: Two Concepts of Causation, in: Collins, J., Hall, N., and Paul, L.A., editors; Counterfactuals and Causation, Cambridge: MIT Press 2004, 225–276

Hawking, S. 1991, Eine kurze Geschichte der Zeit (A brief history of time), Rowohlt, ISBN 3-499-60555-4

Hawking, S.,1998, A Brief History of Time. From Big Bang to Black Holes, Bantam Books, New York

Healey, R., 2012, Quantum Theory: a Pragmatist Approach. British journal of Philosophy of Science, preprint, [2012]

Hume, David, 1740, (Selby-Bigge, L. A. and Nidditch, P. H., editors) A Treatise of Human Nature, Oxford: Clarendon Press 1978

Hume, David, 1748, An Enquiry Concerning Human Understanding, Oxford: Oxford University Press 1999

Ihmig, K.-N., 1999, Ernst Cassirer and the structural conception of objects in modern science: The importance of the "Erlanger Programm", Science in Context 12, 513-529

Iorio, L., 2009, On the recently determined anomalous perihelion precession of Saturn. The Astronomical Journal 137 (2009) 3615-3618

Itzkoff, S.W.,1997, Ernst Cassirer: Scientific Knowledge and the Concept of Man. University of Notre Dame Press [1971]

Irrationality, 2014; (http://en.wikipedia.org/wiki/Irrationality)

Ives, H. E.,1952,: Derivation of the mass-energy relation. In: Journal of the Optical Society of America. 42. 1952, S. 540-543

Jaynes, E. T. ,1957, Information theory and statistical mechanics I   *Physical Review* 106:620;   Information theory and statistical mechanics II   *Physical Review* 108:171.)

Jaffe, R.L., 2005. The Casimir Effect and the Quantum Vacuum, Phys. Rev. D72(2005)021310; arXiv:hep-th/0503158v1, DOI:10.1103/PhysRevD.72.021301

Kant, I., 1958, Critique of Pure Reason (1781). Norman Kemp translator, London, Macmillan

Krenn M., Huber M., Fickler R., Lapkiewicz R., Ramelow S., Zeilinger A., 2014, Generation and Confirmation of a (100x100)-dimensional entangled Quantum System, PNAS; PNAS. DOI: 10.1073/pnas.1402365111).

Kuhn, T.S.,1970, The structure of scientific revolutions, University of Chicago Press

Laloe, F., 2001, Do we really understand quantum mechanics? Am. J. Phys. 69, 655-701
Layzer, D.,1975. "The Arrow of Time", *Scientific American*, 233:56-69.

Layzer, D.,1977. "Information in Cosmology, Physics and Biology", *Int. J. Quantum Chem.* 12 (suppl. 1): 185-95

Layzer, D.,1988. "Growth of Order in the Universe", in: *Entropy, Information, and Evolution: New Perspectives on Physical and Biological Evolution*, ed. B. H. Weber, D.J. Depew, and J.D. Smith, 23-40. MIT Press.

Layzer, David.,1990. *Cosmogenesis: the Growth of Order in the Universe* (section:"Biological Order"). Oxford University Press.).

Layzer, D.,1990, Cosmogenesis,
www.informationphilosopher.com/solutions/scientists/layzer/..

Lewis, D., 2004. How Many Lives Has Schrödinger's Cat? Australasian Journal of Philosophy 82 (1): 3-22

Lichnerowicz, A., 1955, Theories relativistes de la gravitation et de l'elctromagnetisme, Paris, Masson et Cie.

Liu Y. , Liu Ch. Wang D., 2011, Understanding Atmospheric Behaviour in Terms of Entropy: A Review of Applications of the Second Law of Thermodynamics to Meteorology; *Entropy* **2011**, *13*, 211-240; doi:10.3390/e13010211

Ma, X.S., Herbst, T., Scheidl, T., Wang, D., Kropatschek, S., Naylor, W., Wittmann, B., Mech, A., Kofler, J., Anisimova,E., Makarov, V., Jennewein, T., Ursin, R., Zeilinger, A., 2012, Quantum teleportation over 143 kilometres using active feed-forward, Nature 489 (2012) 269-273, DOI:

Malinowski, B., 1948, *Magic, Science and Religion*: Doubleday Anchor Books, Garden City, NY

Marmet, L., 2014, On the interpretation of redshifts: A quantitative comparison of redshift mechanisms,
http://www.marmet.org/cosmology/redshift/mechanisms.pdf

Mathematik - Zitate, 2014:
http://www.arndt-bruenner.de/mathe/Allgemein/zitate.htm

Meyenn, K.v.,1994, Quantenmechanik und Weimarer Republik", p. 48, Vieweg, Braunschweig

Nagel, T. 2012, Mind and Cosmos, Why the Materialist Neo-Darwinian Conception of Nature is Almost Certainly False, Oxford University Press

Naica-Loebell, A., 2001, Interview with A. Zeilinger, 07/05/2001; http://www.heise.de/tp/artikel/7/7550/1.html

Napier, W.M., Guthrie, B.N.G., 1997, Quantized redshift: A status report. J. Astrophys. Astr. (1997) 18, 455–463

NGC, 2014, Hubble_interacting_Galaxy_NGC_5257_(2008-04-24).jpg

Nicolis, G., I. Prigogine, Self-Organization in Nonequilibrium Systems, John Wiley, (1977), 231.

Nucleosynthesis, 2014, http://en.wikipedia.org/wiki/Nucleosynthesis

Ohanian, H. C.; Ruffini, R.; 1994, *Gravitation and Spacetime*, W. W. Norton & Company, ISBN 0-393-96501-5

Olindo de Pretto, 1903; http://en.wikipedia.org/wiki/Olinto_De_Pretto

Paal, G.. 1970. The global structure of the universe and the distribution of quasi-stellar objects. Acta Phys. Acad. Sci. Hung. , 30, 51-54

Paltridge, G.W.,1979, Climate and thermodynamic systems of maximum dissipation. *Nature* 1979, *279*, 630–631.

Pais, A., 1979, *Einstein and the quantum theory*, Reviews of Modern Physics 51, 863-914 (1979), p. 907

Pais, A. 1995, *Ich vertraue auf Intuition*, Spektrum, Heidelberg 1995, p. 196

Penrose, R.,1994, Shadows of the Mind. A Search for the Missing Science of Consciousness, Oxford University Press.

Penrose, R. ,1989, *The Emperor's New Mind: Concerning Computers, Minds, and the Laws of Physics*, Oxford Univ. Press, 1989.

Planck`s Radiation theory, 1897, Chapter III, on irreversible radiation processes, UC Press E-Books Collection,
http://publishing.cdlib.org/ucpressebooks/view?docId=ft4t1nb2gv&chunk.id=d0 e2674&toc.depth=1&brand=ucpress)

Prigogine, I.,1980, From Beginning to Becoming, San Francisco, 1980

Prigogine, I., Stengers, I.,1984, Order out of Chaos, Toronto 1984

Quantum nonlocality, 2014, http://en.wikipedia.org/wiki/Quantum_nonlocality

Popper, K.R. 1979, Die beiden Grundprobleme der Erkenntnistheorie, (Herausgeber: Troels Eggers Hansen), 1979, 426-427, J.C.B. Mohr (Paul Siebeck) Tübingen)

Redshift I, 2014, http://en.wikipedia.org/wiki/Redshift, retrieved 16/05/2014

Redshift II, 2014,
http://hyperphysics.phy-astr.gsu.edu/hbase/astro/redshf.html#c3, retrieved 16/05/2014

Relativity criticism, 2014:
http://en.wikipedia.org/wiki/Criticism_of_the_theory_of_relativity;
http:// www.ekkehard-friebe.de/95yearsrelativity.pdf

Relativity, 2014, http://www.metaresearch.org/cosmology/gps-relativity.asp

Relativity theory, 2014, www.kritik-relativitaetstheorie.de;
http://www.anti-relativity.com/ retrieved 30/01/2014

Ritz, W., *Ann Chim Phys*, 1908, 8(13):145

Ross, J., 2008, *Thermodynamics and Fluctuations Far From Equilibrium* (12.5: Invalidity of the principle of Maximum Entropy production, pgs. 119. Springer.).

Rusu, M. V. and Baican, R. ,2010. Fractal Antenna Applications, Microwave and Millimeter Wave Technologies from Photonic Bandgap Devices to Antenna and Applications, Igor Minin (Ed.), ISBN: 978-953- 7619-66-4, InTech, Available from: http://www.intechopen.com/books/microwave-and-millimeter-wave-technologies-from-photonic-bandgap-devices-to-antenna-and-applications/fractal-antenna-applications)

Ryckman, T. ,1999, Einstein, Cassirer, and General Covariance - Then and Now. Science in Context 12, 585 – 619

Sagittarius A*, http://de.wikipedia.org/wiki/Schwarzes_Loch, retrieved 29/04/2014

Saphiro, R., 1986, A Sceptic`s Guide to the Creation of Life on Earth, New York

Selleri, F. ,1990, Quantum Paradoxes and Physical Reality (ed. Van der Merwe, A.), ISBN 0-7923-0253-2

Schmitz-Regal Ch., 2002, Die Kunst offenen Wissens: Ernst Cassirers Epistemologie und Deutung der modernen Physik. Kassirer – Forschungen, Band 7, Hamburg [2002] Hutchinson, London

Seidl G., 2014, www.geryseidl.at

Sheldrake, R., 1995, A New Science of Life: The Hypothesis of Morphic Resonance, Park Street Press, ISBN-10: 0892815353

Smarandache, F., Yuhua, F.,Fengjuan, Z., 2013: Unsolved problems in special and general relativity, Beijing, www.gallup.unm.edu/~smarandache/UnsolvedProblemsRelativity.pdf

Solar sail, 2014, http://en.wikipedia.org/wiki/Solar_sail

Standard model, 2014, http://en.wikipedia.org/wiki/Standard_Model

Swenson, R. 1997, Advances in Human Ecology, Vol. 6, pp. 1-47, *JAI Press, Inc.* http://rodswenson.com/humaneco.pdf

Szilard, L ,1929, Über die Entropieverminderung in einem thermodynamischen System bei Eingriffen intelligenter Wesen. *Zeitschrift für Physik* 53:840-856

Time paradoxes, 2014, http://en.wikipedia.org/wiki/Temporal_paradox, retrieved 04/2/2014

Tifft, W. G., 1973,. Fine Structure Within the Redshift-Magnitude Correlation for Galaxies. The Formation and Dynamics of Galaxies: Proceedings from IAU Symposium no. 58 held in Canberra, Australia, August 12-15, 1973. Edited by John R. Shakeshaft. International Astronomical Union. Symposium no. 58, Dordrecht; Boston: Reidel, 243)

Toyabe, S., Sagawa, T., Ueda, M., Muneyuki, E., & Sano, M., 2010, Experimental demonstration of information-to-energy conversion and validation of the generalized Jarzynski equality *Nature Physics* DOI: 10.1038/nphys1821

Tributsch H.,2006,: Quantum Paradoxes, Time, and Derivation of Thermodynamic Law: Opportunities from Change of Energy Paradigm. J. for General Philosophy of Science 37, 287-306

Tributsch, H. 1979, Rückkehr zur Sonne, Safari, Berlin

Tributsch, H., 2008, Energy, Time and Consciousness, Shaker Media, Aachen, [2008]

Tributsch, H., 2008, Erde, wohin gehst Du?, Solare Bionik-Strategie: Energie-Zukunft nach dem Vorbild der Natur, Shaker Media, Aachen (2008) 346 Seiten, ISBN 978-3-86858-044-0).

Tributsch, H., 2012, Energy Bionics: The Bio-Analogue Strategy for a Sustainable Energy Future, p. 415 - 464, in "Carbon-Neutral Fuels and Energy Carriers (N.Z. Muradov, T.N. Veziroglu, eds) CRC Press, Taylor & Francis Group (2012), ISBN 978-1-4398-1857-2;

von Neumann, J. (posthumous, [1966]) *Theory of Self-Reproducing Automata*, University of Illinois Press, Urbana

Kohaut, E., Weiss, W.; Das Rätsel Gravitation; Edition Va Bene 2007; ISBN-13: 978-3851671957

Werkmeister, W.H., 1958, Cassirer`s Advance beyond Neo-Kantianism, in P.A. Schilpp (ed). The Philosophy of Ernst Cassirer, The Library of Living Philosophers, 757-798,   Open Court Publishing Company, La Salle, Illinois

Wittgenstein, L.1918, 1922, Tractatus Logico-Philosophicus

Yourgrau, P., 2005 : *Gödel, Einstein und die Folgen: Vermächtnis einer ungewöhnlichen Freundschaft*, C.H.Beck,

Zeilinger, A.,1999,, A foundation principle for quantum mechanics, Foundations of Physics 29, 631

Zeilinger, A., 2000, The quantum centennial, Nature 408, 639.

Zeilinger, A., 2007, Einsteins Spuk, Teleportation und weitere Mysterien der Quantenphysik, Goldmann

Zukav, G., [1979], Dancing Wu Li Masters, An Overview of the New Physics. Rider

Credits:

Fig. 1. - Immanuel Kant within fig. 1: Digitalisat der Universitäts- und Landesbibliothek Halle; http://digital.bibliothek.uni-halle.de/hd/content/structure/1344870;
aus: Zweihundert Bildnisse und Lebensbeschreibungen berühmter deutscher Männer / [Ludwig Bechstein] Leipzig : Wigand, 1857
Fig. 3: picture environmentalist: fotolia, bjul #55525689
Fig. 4: https://de.wikipedia.org/wiki/Ernst_Cassirer#/media/File:Cassirer.jpg
Fig. 5: Niels Bohr: Deutsches Historisches Museum, Berlin
Fig. 22: B. Kaiser, TU Darmstadt
Fig. 30,31,40: NASA
Fig. 43: poultry farming: fotolia, Sergey Bogdanov #65859026
Fig. 43: cattle breeding: fotolia, Ewais #82321072
Fig. 43: fossil energy industry: fotolia, industrieblick #84080535
Portrait Author: B. Baumgartner-Tributsch
All other pictures are from the author.

Cover: Surface pattern of a mushroom growing on hazelnut
Illustrations and Coverdesign: Barbara Baumgartner-Tributsch
Printed by CreateSpace

Abstracts

Abstract 1
By introducing just two logical and entirely justifiable modifications into our present energy concept, the strange irrationalities and paradoxes in fundamental science can be eliminated. As a consequence, the scientific concept of our world would be reshaped from quantum physics to biological evolution and the understanding of our universe. And a new mystery is then revealed. Is development of intelligence and mind the destiny of cosmic evolution?

Abstract 2
It is shown that the counter-intuitive and irrational character of quantum paradoxes may be caused by a lack of information available for the observer on the basis of quantum theory. Since information has an energy value, this means that it should then be the energy concept implemented, the symbolic form, the energy criteria through which nature is understood, which may in part be incomplete for quantum physics. This possibility was investigated and implemented via a "dynamic" energy concept, derivable from the principle of least action, which generates a fundamental arrow of time and is also applicable to quantum phenomena. By correcting the energy formalism accordingly, by additionally considering the missing relevance of space for energy in quantum theory, quantum paradoxes do indeed disappear. This is demonstrated with the double-slit experiment, for non-causality, for quantum correlation and for additional paradoxes. An information self-image of matter, needed for a re-interpretation of the particle-wave dualism of matter, turns out to be very helpful in eliminating irrationalities. It also leads to a local explanation of constant light velocity, and thus makes the assumption of a four-dimensional space-time unnecessary. On this basis, the theories on the universe were re-designed and a new interpretation of gravitation is given. Fundamental natural laws can really be understood rationally and their time-orientation also induces self-organization of information. This allows a materialistic description of nature including mind. Maximum energy turnover, within the given restraints, and development of mind are identified to be the actual aims of evolution.

ABOUT THE AUTHOR

Helmut Tributsch is physicist and a professor for physical chemistry who retired from the Free University, Berlin and the Helmholtz-Centre for Materials and Energy. He specialized in the conversion of sustainable energy and in bio-mimetic energy systems, a subject which he still teaches at the Carinthia University of Applied Sciences in Austria. He has published more than 450 scientific publications and 10 books and now lives on his mountain farm in Friuli, Italy.

www. helmut-tributsch.it

www.ingramcontent.com/pod-product-compliance
Lightning Source LLC
Chambersburg PA
CBHW082304200526
45168CB00018B/3197